Seeing and Being Seen

Aesthetics and Environmental Philosophy

Edited by Joshua Coleman

Hamilton Books

Lanham • Boulder • New York • Toronto • Plymouth, UK

4501 Forbes Boulevard, Suite 200, Lanham, Maryland 20706
Hamilton Books Acquisitions Department (301) 459-3366

Unit A, Whitacre Mews, 26-34 Stannary Street,
London SE11 4AB, United Kingdom

Printed in the United States of America
British Library Cataloguing in Publication Information Available

Library of Congress Control Number: 2017955132
ISBN: 978-0-7618-6995-5 (cloth : alk. paper)—ISBN: 978-0-7618-6996-2 (electronic)

∞™ The paper used in this publication meets the minimum requirements of American National Standard for Information Sciences Permanence of Paper for Printed Library Materials, ANSI/NISO Z39.48-1992.

Contents

Introduction

Joshua Coleman

Part of real and substantive change toward healing the natural environment has to include the way we conceive of nature in the first place. Whether our conceptions are the result of experience or the formative agent of experience (or, most likely, a continual back and forth, the agency in the very relation), there is a connection between how we think toward nature and how we act. Put simply, if we see nature differently, we are more likely to be impacted by the litany of scientific evidence prophesying its demise like a voice in the wilderness.

A recent *New York Magazine* article points to something not given much airtime, but indicates the ecological crisis might be more dire than previously thought. Specifically, the melting of arctic permafrost, not a well-publicized concern until recently, is a reality much worse than rising waters because it contains " . . . 1.8 trillion tons of carbon, more than twice as much as is currently suspended in the earth's atmosphere."[1] The bigger danger is that this carbon could be released as methane, which would have immediate consequences. David Wallace-Wells writes that methane is:

> . . . 34 times as powerful a greenhouse-gas warming blanket as carbon dioxide when judged on the time scale of a century; when judged on the timescale of two decades, it is 86 times as powerful. In other words, we have, trapped in Arctic permafrost, twice as much carbon as is currently wrecking the atmosphere of the planet, all of it scheduled to be released at a date that keeps getting moved up, partially in the form of a gas that multiplies its warming power 86 times over.[2]

As important as it is to publicize these alarming discoveries, it is still necessary to evaluate how we view and conceive of nature for those discoveries to have an impact that will *change the desire* to change our behavior.

Thus the contribution of this collection of essays lies in its attempt to articulate, from a variety of perspectives, the critical import of a philosophical and theological aesthetics of nature, and in turn how that aesthetics impacts our ability to care for nature as an extension of our intrinsic desires. The current title, *Seeing and Being Seen: Aesthetics and Environmental Philosophy*, suggests a relational conception of an aesthetics of nature, in hopes of overcoming subject/object ways of being that lie at the heart of materialist approaches. To that end, an understanding of dwelling in and with nature must be articulated as part of this relational aesthetics.

The uniqueness of the volume lies in the breadth of subject matter covered between articles, which will demonstrate the range of importance with regard to the place of aesthetics, one that impacts everything from ecofeminism to animal rights to human psychology to political theory, and a plethora of issues within these. Part of the goal, then, for aesthetics to take its proper place at the environmental table, means articulating this breadth of relevance while simultaneously overcoming reductionist, materialist presuppositions toward nature, ones which continue to dominate environmental thought to this point.

It is vital, then, to have a working idea of an aesthetics of nature, as well as what is meant by the term "aesthetics" and "nature." Of the many salient points made in his work, *The Noetics of Nature: Environmental Philosophy and the Holy Beauty of the Visible,* contemporary thinker Bruce Foltz suggests that much of our applied aesthetics towards nature comes from conceptions of beauty borrowed from specific cultural and temporal mores, usually from the visual art world. In other words, the aesthetic conceptions normally applied to nature initially arise somewhere else, typically from the dominant approaches to art of a certain period. For this reason, there has been little emphasis on an aesthetics of nature as such, one that could move between cultural contexts and understandings of cultural beauty. This is meant less in terms of *what* is beautiful and concerned more with *how to approach the perception of beauty* in the first place. To this end, groundwork toward an aesthetics of nature is articulated in the first article, "Seeing Nature Whole in Fragments," by contemporary thinker Frank Seeburger. Seeburger's approach will facilitate the great variance of subject matter between the articles in the remainder of the volume.

Because of the current materialist presuppositions in environmental philosophy, the role of aesthetics is likely seen as having marginal concern for a small audience, and with limited impact for practical environmental issues. However, a major goal of this work is to increase attention to a key but overlooked piece of the environmental puzzle by demonstrating the deep impact of aesthetic presuppositions upon the entire panoply of environmental concerns. Far from being marginally relevant, our presupposed aesthetics often (and just as often unconsciously) determines the way we see and value the natural world. Without a closer look at this issue, we are less likely to have the desire to implement necessary changes, whether to personal lifestyle or to public policy.

Seeburger's article, "Seeing Nature Whole in Fragments," addresses a critical aspect of noetic perception not addressed in Foltz's book: Beyond the iconic way of being and seeing into nature (which Foltz articulates), Seeburger explores as well the perception of "being seen," that seeing into nature involves more than action from the seer, even if nature functions as an icon to initially unperceived transcendence, but also involves *becoming* an icon through which nature sees in and through the seer. This two way sense of iconic dwelling, articulated in various forms by Henri Nouwen and Jean-Luc Marion, adds a critical piece to the discussion, that only through such receptive action does one come to truly know nature, by way of "dwelling" amidst the vulnerability of "being known." Moreover, it is necessary to allow this "being seen" if we are to truly overcome subject/object ways of engagement.[3]

The following chapter, "The Noetics of Poverty in the Paintings of Vincent van Gogh," develops from the paper I presented (inspired largely by Foltz's book) to the 2014 Southwest American Academy of Religion. Since that time the article grew significantly through collaboration with thinker and writer J. Porter, with whom I co-authored the final chapter as well. In chapter two we discuss the place of "noetics" as a traditional antiquarian concept of perception in and of nature, that there is a knowledge that comes by way of participation, and how such perception proves analogous to the way in which painter Vincent van Gogh articulated his depictions of the poor. In distinction from a cultural aesthetic put upon him by the rich, one that opposed nature in its aesthetic approach, Van Gogh's "aesthetics of nature" with the poor signaled an entirely altered strategy, in a way that forced the viewer to participate in the painting and thus we call it "noetic;" a term which, again, describes a type of knowledge as the end of a process whereby the viewer is not objective observer, but actively participates in the viewed and comes to knowledge through such participation.

This approach went against the dominant cultural aesthetic of the rich who did not call for such participation, and more often objectified the viewed by preferring brightness of color rarely found in nature. Van Gogh's impressionism better captures an aesthetics of nature as relational, giving and receiving, because it calls the viewer into an active piecing together of the viewed, as well as the *reception of being viewed.* This means the viewer's gaze does not merely grasp an object which may or may not literally mimic nature, but rather, in piecing together the scene herself, becomes a lover of nature by *being with* nature through this active and imaginative process.

The first two chapters provide a launching point for Dr. Mark Flory, who focuses on specific practices that aid in this same perceptive process by taking a closer look at the Eastern Christian tradition of Hesychasm and specifically the thought of Medieval Greek monk, Gregory Palamas. Drawing from Plato's cave allegory and Kierkegaard's stages of existence, Flory argues (among other things) that just as the cave is no longer the same for the one who has left and returns, and just as the aesthetic and the ethical stages are completely transformed once one *is leaped* (to use Flory's logic) into Kierkegaard's religious stage, the tradition of Hesychasm is one which has the practitioner enter an horizon of practices that are by their very nature transforming; moreover, they transform both the practitioner *and* the cosmos.

Critically, there is a sort of empiricism at play in Flory's investigation of "Hesychasm" and the ensuing transformation of the perceptive capacities found therein. There are certain a priori assumptions in such a practice, whether it is about cosmology or anthropology, and those assumptions have everything to do with the practical experience of transformation, just as our a priori assumptions about nature have everything to do with our experiences and desires toward nature. Analogously, contemporary science does not avoid having a priori assumptions that certainly impact what can be known by the modern empirical method; as well an ethic, even if the ethic is the assumption that the observer is not to be swayed by ethical considerations, as just one example.

Yet Hesychasm is not primarily about experience, given the very important role ethics has on the transformation of the hesychast practitioner. This means the practice is not strictly mechanical, and it is ultimately about the continual transformation of practitioner and the world around her. This transformed perception exercises what are called "the spiritual senses" in Palamite, patristic jargon, enabling those who develop them to dwell differently within the natural world. Flory suggests that the relation to nature can change so drastically that what one perceives to be nature changes as well; in overcoming subject/object being, relational knowing leaves nothing static or the same, and one's perception of beauty is radically altered.

From there, Dr. Donald Turner's article, "Rethinking Animality, Humanity and Divinity, With Inspiration From George Bataille," will continue to investigate aesthetics with specific application of Bataille's study of those cultures that perceived divinity in animal life, in distinction from one all too common interpretation of the Judeo-Christian Scripture as justifying animals predominantly for use, or that it is only humans within whom the divine resides. Dr. Turner demonstrates how our aesthetics impacts our language; we eat "meat," not cows; "pork" and not pigs, and that both have bearing upon our ability to see and be seen by animal life. He achieves this through an analysis of the intimacy and religious respect certain cultures had and gave to the animals who provided them sustenance, particularly those cultures which depicted icons of these animals and saw, according to Bataille, as well as the American Navajo, divinity within in them. Turner provides practical specifics to help rethink our present day relationship in a way that can alter human behavior and impact the environment as a result. Of his many goals, Turner aims to overcome the system of bifurcation which creates a largely unseen environmental burden wherein consumer habits remain obscured to the consumer, as well as the living welfare of the animal now divested of its life to be dinner.

Dr. Turner notes the aesthetic consciousness that protects the Bald Eagle as a symbol of freedom and strength, unlike a different bird which endures severe maltreatment and whose comb is often processed into anti-wrinkle creams, demonstrating a somewhat narcissistic aesthetic, one that projects strength and nobility onto one animal, while dismissively torturing another which might not reflect the dominant cultural aesthetic. In this case, the aesthetic impacts which response is compassionate and which leads to an utter void of compassion if not outright aggression. The applied conception of our aesthetics, combined with our ability/inability to see the sacred through that conception, directly impacts the care and respect that goes into our relationship to non-human animals, our ability to dwell with fragments of nature implicit throughout.

Dr. Sharon Adams picks up where Seeburger began the book to articulate a Nietzschean aesthetics that transcends binary oppositions. Taking this notion that philosophy has often focused on totalities, and that part of the antidote to them is one sort of Nietzschean, Eco-Feminist approach, Dr. Adams looks at Nietzsche, Lou Andreas Salome and Goethe (among others) to give voice to the "gaps" between binaries that reveal the binaries not merely as fraud, but as keeping us from giving ourselves (as fragment) to the "fragment" in front of us. Put simply, the paradoxical tragedy (as well the hubris) of totalizing thought is that we cannot conceive of its death, and yet every fragment, from person to plant, will in fact die. This means the very attempt to systematize the whole is a perilous fantasy not based in lived reality, offering us little in terms of what we can do to better enter and care for the environment in front of us. We cannot conceive of the whole dying because we cannot conceive of the whole in the first place as having reference in lived experience, the latter of which always and everywhere decomposes. As such, totalizing conceptualizations, when put upon nature, lull us into thinking nature will not die "as a whole."

The final chapter seeks to explore the relationship between eros and beauty as they facilitate our collective response to previous environmental trauma. Here it is put forth that the only authentic way forward is to experience what we have done at some level; in the words of Frank Seeburger, to "let trauma traumatize." In order to successfully move away from totalizing philosophical narratives with regard to nature, and by way of letting trauma traumatize, we see a different way of being toward the fragments, disclosed by the unconscious which, in giving us shards of symbols and incomplete narratives, leads us through the very journey which might have us change. The conscious mind plays its part in filling in the narratives, but can also be our greatest hindrance, as it tends to tell sanitized, tidied tales as an attempt to self-protect and turn away from the often painful processing of the unconscious.

To demonstrate this, J. Porter and I follow closely the work of contemporary scholar Avivah Zornberg, looking at the role the unconscious plays in overcoming idolatries of all sorts, with the implication that, when we appropriate nature as sacred, we have a unique opportunity to stop seeing it through a materialist lens. This process, however, as Zornberg herself suggests, is an "erotic" one, taking us outside ourselves to journey through the desert of the unconscious. Yet this is the place we become participants with nature and the divine; this is the painful sickness which Moses held in his bones in order to have God "change His mind" with regard to destroying creation after Israel's own idolatry. Once again, according to the narrative (and perhaps to many a scholastic theologian's chagrin), everything changes in the story, *even God's mind.* We might be tempted to over-interpret this to make God fit the scholastic imperative that God never changes, and that it is Moses' perception of God and so on. Yet in doing so we would miss a critical opportunity because in striving for rational consistency we'd risk making God once again into that which does not change; i.e., an idol, and fail to overcome subject/object being.

The question then becomes, how do we allow beauty to unsettle us while not trying to control it, to let any healing with regard to our relationship to nature result from this internal wandering, one which calls us outside of ourselves to engage and be changed in unpredictable and unforeseen ways? The ability to do this is part of purging the materialist idolatry of the natural world and its totalizing narratives which have led to so much obvious destruction in the last two centuries. There is a way forward, however, if we are willing, and it must include rethinking our a priori aesthetics as an essential piece of the puzzle.

This volume should allow a student of Philosophy, Theology or Environmental studies to get a sense of the interrelatedness of crucial environmental preconceptions and a wide array of issues impacted by those preconceptions, ones we all too often carry around with us unconsciously. It is then hoped that further discussions and collective effort to engage a crisis that effects all of us can continually begin again with the expectation of new life, not wearied or dismayed by old narratives that continually fall on deaf ears. May our senses be opened to creative ways of communal change and "may those with ears hear!"

I am grateful to all the contributors of this work. I appreciate your patience with changing deadlines and long dead periods of correspondence. It is with such pleasure that I was able to collaborate with each one of you. I certainly gained far more than I was able to give. Thank you.

NOTES

1. Wells, David Wallace. *New York Magazine,* July 9, 2017.
2. *Ibid.*
3. Thomas Merton remarks on a chapter in William Faulkner's novel, *As I Lay Dying,* entitled, "The Bear." At one point during the hunt, and though they do not see "Old Ben," they are aware that the bear sees them. For Merton, this was indicative of monastic mysticism, a cultivated awareness of "being seen" in and through nature.

Chapter One

Seeing Nature Whole in Fragments

Frank Seeburger

The eye through which I see God is the same eye through which God sees me; my eye and God's eye are one eye, one seeing, one knowing, one love. —Meister Eckhardt

Earth's crammed with heaven,
And every common bush afire with God:
But only he who sees takes off his shoes.
—Elizabeth Barrett Browning

PRELUDE: BEHOLDING NATURE: THE AESTHETICS OF NATURE AND THE NATURE OF AESTHETICS

Both phrases, *the aesthetics of nature* and *the nature of aesthetics*, are ambiguous. In both cases, the ambiguity hinges on how one hears the preposition, the *of*: on whether one hears it as an "objective" genitive, or as "subjective" one. Most commonly, in the former expression (*the aesthetics of nature*) the genitive is heard as "objective," so that it means "aesthetics" directing itself towards "nature" as the object of the aesthetic gaze, as it were. On the other hand, the latter expression (*the nature of aesthetics*) is probably most commonly heard as a "subjective" genitive, so that it refers to the "nature"—as when we speak, for example, of "the nature of the state," or "the nature of pigeons"—that belongs to "aesthetics" (taken as a distinctive branch of philosophy for example).

Differently tuned ears, however, might hear both phrases differently. Instead of being heard as the nature that belongs to aesthetics, *the nature of aesthetics* might be heard to mean what nature itself is taken to be, when it is made the object of aesthetics, as the latter has traditionally been understood.

That is, the phrase might be taken to point to what is made of nature—what happens to it, what changes it is subjected to, forced to undergo—when nature is taken as the focus of what is ordinarily thought of as a purely "aesthetic" concern: nature as it is manifest to be in "aesthetics."

In turn, the second phrase, *the aesthetics of nature*, might be heard as a "subjective" genitive rather than an "objective" one, so that it would refer to the aesthetics that nature itself practices, so to speak. The phrase would then be heard to mean how things manifest themselves to *nature's* "aesthetic gaze" as it were.

In the essay that follows, I play with such possibilities of interpretation, in hopes of granting to nature what is due to nature, and to aesthetics what is due to aesthetics. Above all, I play with the possibility that by letting nature itself cease being merely the object of the aesthetic gaze, and allowing nature to become, instead, the one doing the gazing, aesthetics itself might undergo a radical change of nature—and in the process become much more "nature friendly," as I might put it.

To end this "prelude," I want to sound the cord of yet another ambiguity in yet another word, an ambiguity based upon the roots of the word, even though it is not commonly heard there. The word is *behold*, which ordinarily means to gaze upon, of course. However, we might hear the *hold* in *be-hold* differently, the way we do when we speak for example of "holding" our loved ones dear, which is to say "cherishing" them: thereby using "*be*-holding" in the sense of "*up*-holding," in effect. What is more, we might at the same time hear *behold* along the lines suggested by speaking of being *beholden* to someone, thereby meaning "indebted' to them, "owing" them for favors with which they have favored us.

What I hope to suggest through the following essay is that in beholding nature in a way that honors nature's own nature, we may come suddenly to experience nature beholding us back, and to realize how beholden we are to it, how obligated to be-hold it, just as it has so long (from birth, in fact) beheld us.

That would indeed be an "ecological aesthetics" to behold!

1.

In 1929 Martin Heidegger delivered "What Is Metaphysics?" as his official inaugural lecture to assume the chair in philosophy at the University of Freiburg from which his own mentor Edmund Husserl had just retired (Heidegger 1998). At one point in that lecture, Heidegger distinguishes between "the whole of beings" (*das Ganze des Seienden*) on one hand, and "being in the whole" (*das Seiende im Ganzen*). The former, the whole *of* beings, would be everything that in any sense "is"—from gods and angels, to unicorns, vampires, and even such fictions as Martian aliens or golden mountains, to concepts, relationships, and numbers, to aardvarks and Zimbabweans, or even the smallest grains of sand—all put together to form one single totality. In contrast, the latter, being *in* the whole would be everything (again, in any sense of "thing" at all) that *is* (in any sense of "is").

The former (*das Ganze des Seienden,* "the whole *of* beings") is never given to us, Heidegger points out, but the latter (*das Seiende im Ganzen,* "being *in* the whole") is given to us constantly. Being *in* the whole is given to us whenever any single being at all is given to us.

Part of what Heidegger said in "What Is Metaphysics?" had already been said clearly enough almost a century and a half before, by Kant. In the *Critique of Pure Reason* Kant had already said the first half of what Heidegger said in 1929. In that first of his three *Critiques*, Kant had already said that what Heidegger in "What Is Metaphysics?" calls "the whole of being"—"the *totality* of beings," that is, the totality of all that in any sense "is"—is never given to us.

In contrast to Heidegger, Kant used the term *world* to mean such a totality of everything that can in any sense be said to be, and insisted that the "world" so conceived is not even a "possible object of experience." Instead, he assigned to the world so conceived the status of a "transcendental Idea," a guiding and governing idea for thought and inquiry, one which serves the purely heuristic purpose of keeping thought and inquiry open and ongoing, without end. As such a Kantian "transcendental Idea," the world is, in effect, a star that can never be reached, but by orientation to which all the efforts of natural science are always to navigate, in the never-ending voyage of research.

In the famous first line of his *Logisch-Philosophische Abhandlung*, which was first published in 1921, just a few years before Heidegger gave his talk—and which simply means "Logico-Philosophical Treatise," but was brought out in English under the Latin title of *Tractatus Logico-Philosophicus* (Wittgenstein 1922)—Ludwig Wittgenstein wrote: "The world is everything that is the case." ("*Die Welt is alles, was der fall ist.*") Well, both Kant long before Wittgenstein and Heidegger not long after him taught that precisely such a Wittgensteinian world, the world in the sense of the whole, the totality, of "everything that is the case," is never given to us, never itself brought to stand before us as some object (Kant) or singular being (Heidegger).

Contemporary French philosopher Alain Badiou continues to argue today—as he has for many years now, including in his two major works *Being and Event* (Badiou 2005) and *Logics of Worlds* (Badiou 2009)—that not only are we never *given* the whole of being. Rather, he insists, there simply *is* no such whole that we ever even conceivably *might* be "given." By Badiou's argument, there just is no being that would actually *be* the whole, in the sense of the totality, of all beings—no such actual thing as *the* world at all, in

Wittgenstein's sense of that word, whereby it would mean the totality of "everything that is the case."

Badiou uses set-theory to establish his point: a *reductio ad absurdam* on the thesis that there could ever be a "set of all sets." But the same point is already implied in Kant. It applies, in fact, to any and all of what Kant calls "transcendental Ideas," including the idea of the world as the totality of what is.

In the first *Critique* and after, Kant himself argued that it is only within the bounds of possible experience that any of our terms have signifying or "constitutive" meaning at all—that is, pick out aspects of even *possible* "objects of experience." Only within such bounds do those terms give us eyes, in effect, through which we can see anything possible, actual, or necessary—a way of putting the point that accords well with Kant's famous line that "concepts without percepts are empty," just as "percepts without concepts are blind." It is the job, as it were, of concepts, and the words that signify them, to give us windows that open upon possible objects of experience. If they fail to do that job, they lose *constitutive* or *signifying* sense, and retain at most only a *regulative* one—providing direction to ongoing inquiry, as already explained above. If we keep reason itself within its proper bounds, according to Kant, then we refrain from applying our concepts and terms beyond their purely "phenomenal" sense, their sense for signifying possible objects of experience.

Well, insofar as the "world," in Kant's and Wittgenstein's shared sense of "the totality of whatever 'is'," is not even a possible object of experience, as Kant has it, then to keep reason within its proper bounds one thing we can never attribute to such a totality is "existence" itself, whether possible, actual, or necessary. Strictly speaking, we do not even know "what in the world or out of it, for that matter"—to borrow a way of speaking from Kant himself—could be *meant* by attributing "existence" to such a totality, since "existence" itself is precisely one of the concepts that Kant explicitly argues *has* meaning only when applied *within* the bounds of "possible objects of experience." (The same reasoning applies to another of Kant's pure Ideas, by the way: the idea of God as a being "transcendent" to the world. The reader should keep that issue in mind, though I will not address it explicitly in this essay.)

There just is no such thing, either in or under the heavens, as what Kant and Wittgenstein call the "world." To speak as Heidegger does, there just is no such thing as *das Ganze des Seienden*, no such thing as the whole *of* being, of all that "is."

2.

Yet each time we are given anything whatever, in whatever way, then we are given it as a being, as something that "is," *in* the whole. Whether awake or dreaming, perceiving or imagining, seeing or hallucinating, we are always everywhere, so far as we are given anything at all, given Heidegger's "being in the whole," *das Seiende im Ganzen*.

It is with regard to "being as such and in the whole"—which is that about which what Aristotle called "first philosophy" or "theology," and which long ago came to be called "metaphysics," is the inquiry (as Heidegger never tires of reminding us)—that we must say what contemporary French linguist and philosopher Jean-Claude Milner has recently argued should be said of literature and, in general, works of art and thought. In his recent book *La puissance du detail: phrases célèbres et fragments en philosophy* ("The Power of Detail: Celebrated Phrases and Fragments in Philosophy"), Milner himself celebrates the fragment as such, and its libratory power.

Milner contrasts the power proper to the fragment with the power that seeks everywhere always to bring everything back into relation to some whole or totality. Speaking of the dominance of such a totalizing strategy, he writes (Milner 2014, 10)[1] :

> Totalities have long been privileged. It has been proclaimed that it is impossible to understand an isolated bit of discourse if one does not reintegrate it into its whole context. Impossible to stay with an isolated discursive bit without completely retracing the figure within which it inscribes itself, the oppositions of which it is a term, the genealogy of which it is a moment, because in the last instance, it goes without saying, the true, that's the totality. Against such a bias, there's nothing to object. But one can make other choices.

Milner immediately goes on to characterize one such "other choice," embodied above all in works of *art:*

> That [. . .] the *Aeneid* as a whole clarifies the significance of a verse of Vergil—certainly; but that's not what's essential. The effect of meaning is born from the phrase itself, from the few words it gathers together, from its being cut out, from its incompleteness. The contexts that one restores, the circumstances that one recalls, they all count only to the degree that, thanks to them, what's fragmentary is perceived as fragmentary, the opacity as opacity, the equivocation as equivocation. The interpretation should not seek to reestablish the easy continuity of the fragment within its place of origin, but rather to make manifest the strokes of the chisel that have torn it from that place.

It is only such a fragmentizing, as opposed to totalizing, reading and analysis that works with the materials that constitute a "culture" that, as Milner has already written just before the passages cited above, is "able to incite rebelliousness" ("*insoumission*")—the very resistance that can break us out of the totalizing strategy that otherwise entraps us (Milner 2014, 9). To develop our capacity to read and analyze in such a rebellious spirit is, according to Milner, a "duty" today:

> To exercise such acuity, to intensify it, extend it, is a duty. I'd speak of ethics, if I weren't sure to be misunderstood. Out of prudence, I will instead speak of political duty. Vigilance is often invoked; it serves nothing, if it is not accompanied by hyperesthesia. When it's a matter of intention and conduct, it doesn't suffice passively to receive, it's necessary to discern the details. It doesn't suffice passively to receive the materials, it's necessary to treat them actively. Culture goes to ruin, if one leaves it in the state in which one finds it. Culture or rather cultures, diverse in their languages and objects, are speaking media that one must work with. Only then do they yield their hoped for benefits. One who treats discourse as matter to be handled understands and sees the realities better. One who understands and sees the realities is less easy to domesticate. Contrary to current opinion, I maintain that cultures are able to incite rebellion. From that point of view, they all share the same virtues, as different as they may be [. . .].

A few pages later Milner adds that that is the difference between "cultures" and "knowledges" or "bodies of knowledge" ("*saviors*")—which includes, of course, mathematics, the sciences, and technologies. Considered as discourse, a body of knowledge (as the Wittgenstein of the *Tractatus* also clearly saw and understood) consists of systems of true propositions organized under a totalizing perspective (Milner 2014, 11).

In contrast, for Milner the discourse that constitutes a culture really consists of nothing but fragments. A culture is "no more and no less than a dustbin of phrases and sayings, citations, adages, maxims, clichés," as such suited to "facilitate rather than block the irruption of new ways of thinking." The worker in culture, as opposed to the worker in the knowledge industry, is one who strives always anew to tear fragments and details loose from their moorings, and set them free. To build culture as opposed to knowledge, what counts is not taking the part of the whole. What counts is not sharing the bias (in French, *parti pris*, from *parti* "camp, group, party," + the past participle of *prendre* "to take or grasp") for the totality. What counts is, rather, "to take the part of the fragment and take it resolutely." Thus: "If the fragment is well known, return to it its strangeness. If it passes for limpid, bring back its obscurities. If it is supposed to close the debate, split it open at the middle to uncover its hidden uncertainty. [. . .] Going against the all-powerfulness of habit and memory, undertake at each instant to disclose the in-habitual and the forgotten."

From such passages begins to emerge a vision of culture—or, rather, a way of seeing culture*s*, in the plural, since there is only the multiplicity of cultures, and no such thing as "culture itself" as some sort of totality, any more than there is any totality of what is—as itself essentially *fragmenting*. It is not only that cultures always and only come to us, when they really come, as isolated "fragments," "details," "installations," "interventions," and the like. It is that cultures' work—their work of cultivating, en-culturating worlds—requires making fractures, breaks, gaps, and furrows in what otherwise would lay claim to being the whole, the *totality*, of what is, rather than just one more being, one more fragment, *in* the whole.

With regard to *art* (the master-builder of cultures, as it were) it is as though the work of art fails at its work whenever it fails to shatter itself apart into fragments. That is, to turn the thought around, it is as if it is only in its fragmentation that the work of art can work as art, and not degenerate into cliché, mere window-dressing and divertissement—sheer artifice.

Rainer Maria Rilke's "*Archäischer Torso Apollos*" ("Archaic Torso of Apollo"), a poem addressing all that remains of an ancient statue of Apollo from which the head and limbs have been broken off and lost long ago, testifies to this power of the sheer fragment of a work of art to do art's real work--and how pure art, freed from all artifice, works precisely in and as such fragments. Here is the poem, first in the original German, then in Stephen Mitchell's English translation (Rilke 1995):

Wir kannten nicht sein unerhörtes Haupt,
darin die Augenäpfel reiften. Aber
sein Torso glüht noch wie ein Kandelaber,
in dem sein Schauen, nur zurückgeschraubt,

sich hält und glänzt. Sonst könnte nicht der Bug
der Brust dich blenden, und im leisen Drehen
der Lenden könnte nicht ein Lächeln gehen
zu jener Mitte, die die Zeugung trug.

Sonst stünde dieser Stein entstellt und kurz
unter der Schultern durchsichtigem Sturz
und flimmerte nicht so wie Raubtierfelle

und bräche nicht aus allen seinen Rändern
aus wie ein Stern: denn da ist keine Stelle,
die dich nicht sieht. Du mußt dein Leben ändern.

We cannot know his legendary head
with eyes like ripening fruit. And yet his torso
is still suffused with brilliance from inside,
like a lamp, in which his gaze, now turned to low,

gleams in all its power. Otherwise

the curved breast could not dazzle you so, nor could
a smile run through the placid hips and thighs
to that dark center where procreation flared.

Otherwise this stone would seem defaced
beneath the translucent cascade of the shoulders
and would not glisten like a wild beast's fur:

would not, from all the borders of itself,
burst like a star: for here there is no place
that does not see you. You must change your life.

Whenever the whole of being is given all the power, it overpowers every being in the whole. When that happens, it is our duty to rebel. Art—real art, and not just some hurdy-gurdy sort of entertainment to give everyone "their little poison for the day, and their little poison for the night," as Nietzsche puts it in his "Prologue" to *Thus Spoke Zarathustra*," from Zarathustra's great speech about the "Overman" and the "Last Man"—is there to call us to do our duty.

3.

In the winter semester of 1941–1942—twelve years after he had first delivered his lecture on "What Is Metaphysics?"—Heidegger held a course at the University of Freiburg (Heidegger 2008). It was one of his regular "seminar exercises"(*Seminar Übungen*) and he gave it the title "Practice in Philosophical Thinking." The term I have rendered as "practice" is *Einübung*. Especially as Heidegger uses it for his course, that term indicates, in effect, an "introductory exercise," one designed to lead beginners into a practice. Other translations besides "practice" are "training" or "rehearsal" (as in theatrical or musical rehearsals). At any rate, what is meant is this: learning how to do something by doing it, rather than by talking, reading, or thinking "about" it.

Accordingly, Heidegger begins his course by marking that very difference, the difference between learning "about" some discipline, on the one hand, and actually learning that discipline, on the other. He begins more than one of his courses by marking that same distinction in some fashion. The way he does so in his 1941–42 "seminar exercise" in "philosophical thinking" is a way that he repeats most famously a decade later in his 1951–1952 lecture course, *Was heißt Denken?* (translated into English as *What Is Called Thinking?*).

To learn a discipline, and not just learn about it, one has to *practice* that very discipline itself. As I used to say to students in my own classes, throughout my forty-five years of university teaching, learning a discipline such as philosophy is like learning swimming: no matter how much one learns "about" swimming, through reading or watching films or listening to instructions from swimmers and watching people swim, one will never to learn how to swim without getting in the water and starting actually to do just that, *swim*. If parents want to teach their children how to swim, I'd say to my students, then sooner or later they have to throw their kids in the water and let them swim.

That's how Heidegger himself taught "philosophical thinking," at any rate, not just in his 1941–1942 seminar exercise but also in all his other courses: by taking his students with him into the waters of philosophical thinking itself, and helping them to begin finding their own way to stay afloat in those waters. He didn't spend much time talking *about* philosophy with them before taking them into philosophical waters. Instead, he just led them into the beginners' end of the pool, and then gave them pointers as they began to swim along with him.

One crucially important pointer he repeatedly gave his students in philosophical thinking, whether in 1941–42 or any other year, was to be attentive to what a beginner he, Heidegger, himself remained. In fact, as he says at one point in his 1941–1942 seminar "practice in philosophical thinking" (Heidegger 2008, 154), they should keep in mind as they learn philosophical thinking that "[p]hilosophical exercises are *always* 'beginners' exercises' " (emphasis added). In an expansion appended to that remark he clarifies (Heidegger 2008, 244) that "[i]n such thinking, we are always 'beginners'," since philosophical thinking is a thinking that itself "has to do with the beginning." For that very reason, he warns his students not to look to him as some great "authority" on the matter. Indeed, when it comes to philosophy, he says, "there is, strictly speaking, [. . .] no 'authority'." There are always only beginners where the very business, as it were, is just that: to begin.

Accordingly, instead of informing his students at length about what it was like to think philosophically, he right away threw them into the philosophical water, to begin *doing* some "philosophical thinking" along with him. As was typical in his courses, in his "seminar exercise" in philosophy of 1941–1942 he did that by inviting his students to think along with him as he himself thought along with his predecessors, those beginning thinkers who began to think before him. Specifically, for his exercise class of that semester he invited his students to think along with him as he tried to think along with two of those earlier beginners, two thinkers whom the tradition had handed down as being pretty good at beginning thinking, as it were: Heraclitus, a beginning thinker from way back at the beginning of the very tradition at issue (the tradition called "philosophy"); and Nietzsche, a beginner at the end of that same tradition.

Thus, Heidegger and his students in performing their beginners' exercises in philosophical thinking do not even try to address themselves to the whole of what has come down to us as "philosophy" (no whole of philosophers, *Ganze der Philosophen,* in effect). They instead confine themselves to just two figures from that tradition, two fragments of that whole (two philosophers among the whole, ongoing mass of them, two *Philosophen im Ganze*, as it were). Heidegger invites his students to think along with him as he thinks along with those two philosophers, fragmentary as such a small sampling may be.

What is more, in thinking along with those two fragments of philosophy, Heraclitus and Nietzsche, Heidegger and his students will not even make an attempt to deal with the thinking of either philosopher himself as a whole. Rather, they will deal only with fragments of the thought of each: just a single fragment from each.

"From Heraclitus," Heidegger reminds his students (Heidegger 2008,153), that's all we have to begin with, since "only fragments have been handed down" to us. Heidegger immediately adds that we could almost say it is "by luck" ("*zum Glück*") that we only have such fragments. And as for Nietzsche, given his insistence on writing in aphorisms, we might well say that all he ever *wrote* was "fragments." That too, we could well add, is fortunate, a bit of luck—*good* luck.

"Whether a work [such as Heraclitus's thought as a whole] has survived intact or not," as Heidegger goes on to say, "does not decide over the foundation and the sort of the work's so called 'effect'," precisely because "the 'effect' of a thinker is of its own unique sort." The greatness of a work and its working is not a matter, Heidegger observes, of how large a number of people it has had an "effect" upon, in the usual way of taking the expression that something has had an effect on someone: struck them as "interesting," or "memorable," or the like—as even a cheap but gaudy piece of artificial jewelry can do, for example.

For Heidegger and philosophical thinking as a whole, in fact, what Milner says of literature and Rilke says of an archaic torso of Apollo—and by implication both say of all works of art—it is precisely the fragment that speaks most loudly and clearly. It is only the fragment, in fact, that can say everything.

4.

In 1960 the German language poet Paul Celan was awarded that year's Georg Büchner Prize for Literature. Büchner, to honor whose memory the prize is given, was an important nineteenth century German dramatist and revolutionary activist who died of typhus in 1837, when he was only 23. His great promise was thus cut short, and he left behind only a fragment of the contribution to literature he might have made had he lived a normally long life. Nevertheless, despite constituting only a fragment of the whole of which he was capable—or for that very reason, perhaps—his work had a significant impact on subsequent German (and, subsequently, world) literature, and continues to have great power even today, almost two centuries later.

Paul Celan, the eventual recipient of the prize later named after Büchner, was born Paul Antschel to a Jewish family in what was then Romania. As a young man, he was imprisoned by the Nazis, but managed to survive the Holocaust—at least in the ordinary sense of "surviving" it, since he killed himself by jumping in the Seine River in Paris in 1970, when he was only 49.

After being liberated from the Nazi camp-system at the end of World War II, Paul Antschel transposed the syllables of his last name and, relying on the Romanian pronunciation of *c* as *tch*, he took the penname of "Paul Celan." He is perhaps the greatest post-Wold War II German language poet, and certainly one of the greatest of all *post-Holocaust* poets. Indeed, his mature poetry virtually *defines* what it means to write poetry that is truly *after* the Holocaust, "after" it not just chronologically, but in its very essence as poetry. In fact, in his 1960 Büchner Prize acceptance speech *Der Meridian* ("The Meridian"), Celan himself articulates the idea that such truly post-Holocaust poetry as his is actually the very distillation of poetry as such down to its pure essence, what he calls "the pure poem" (Celan 1961).

What Celan calls the pure poem is what is sheer poem, with nothing other. The discourse that makes up the pure poem is pure poetic discourse, which lets be seen and, in letting be seen, liberates. It calls attention to itself only in order to hurl that attention beyond the poem itself, to what it says. Such poetic discourse is discourse that has become just discourse and no more, with no pontification, hullabaloo, prettiness, eloquence, or gaudiness about it. It is discourse without any artifice.

The discourse of the pure poem stands in contrast to the sort of discourse all too typical of politics and politicians, to give a prime example of a discourse filled with itself and its own importance. It is just such self-inflating, counter-poetic, political discourse that one finds dramatically depicted in *Danton's Death*, Georg Büchner's own great historical play about the events of the French Revolution, events that surrounded the one named in the play's title. Celan cites and builds upon that very play, *Danton's Death*, in "The Meridian," his own Büchner Prize acceptance speech.

Such pretentious discourse as makes up the speeches of politicians is not only filled with itself. It also attempts to fill the whole field of public discourse as such, blotting out all voices other than its own. It is a totalizing discourse, one that strives to reduce all to one—a self-important, bombastic, and totalizing discourse that over-fills itself with discursive content, striving to leave no room for anything else to voice itself, or be heard.

Such bombastic discourse, the discourse that is all too common in politics, calls attention to itself by intentional showiness in its drive to totalize. In contrast, pure *poetic* discourse, that discourse which strives to become pure poem is, as such, very plain. The pure poem is discourse plain beyond measure. Political rhetoric puffs itself up pridefully, calling attention to itself for its own sake, trying to drown out all voices but its own. But discourse in the pure poem only calls attention to itself for the sake of vanishing altogether before that to which it wants to give voice.

The discourse typical of politics might be called an "additive" discourse, one that is never at a loss for words and is always ready to add more to what it has already said. Such discourse tends by nature to puff itself up expansively. In contrast, purely poetic discourse could be said to be "subtractive," and to tend naturally toward ever more condensed sparseness. It operates not by joining things together, reconciling them into ever more exhaustive and larger wholes, but by disjoining things, splitting them apart (literally *parting* them) to let each thing stand clear and distinct in relation to all others, all things together in stark confrontation (that is, "face-to-face with" one another—from Latin *con-*, "with," + *frons*, "forehead").

At one point in "The Meridian," Celan remarks that "the poem today shows, as cannot help but be recognized, a strong tendency toward holding its tongue." The German term I've translated as "holding one's tongue" is *Verstummen*, which means to keep quiet, stop talking, become mute (*stumm*). The very tendency Celan addresses in this remark is highly evident in his own poetry.

In following that tendency toward muteness, the poem today "asserts itself," Celan goes on today, "at the fringes of itself, to be able to stand [. . . as] actualized language, set free under the sign of an individuation that is certainly radicalized, but at the same time mindful of the limits imposed upon it, the possibilities opened to it, by language."

Thus, a radical linguistic individuation is at work in poetry that comes truly after the Holocaust, an individuation that pushes language to its extreme limits. Such linguistic individuation to and at the extreme limits of language still remains mindful of its roots *in* language, and is thus *radical* in the etymologically original sense of "going to the root."

Thus, purely poetic, individuating discourse is radical in the double sense of being both "extreme" or "going to the limit" (as accords with common contemporary usage of the term *radical*), and "going to the root" (in accordance with the words own etymological roots). To put the two together: Such purely individuating discourse is radical in going to the limits of language by going to the very roots of language. As such, it is an individuation that brooks no compromise.

Such doubly radical linguistic individuation cannot be found in the chatter of the chatterboxes, including the grandiloquence of the flashiest orators, the fanciest writers. It "can only be found," Celan says, "in the work of poets who do not forget that they speak from the corner of inclination of their own existence, the center of inclination of their own creatureliness."

So conceived, the poem becomes indeed "the language come-to-form of a singular individual [*eines Eizelnen*]—and, in accordance with its innermost essence, presence and the present." Thus, Celan continues:

> The poem is solitary. It is solitary and underway. The one who writes it, comes given along with it.
>
> But doesn't the poem precisely thereby also stand, here, in an encounter—in the mystery of an encounter?
>
> The poem wants others. It needs others. It needs neighbors. It seeks them out. It speaks itself to them.
>
> To the poem, which counts on the other, every thing, every person is a form of this other.
>
> The attention [*Aufmerksamkeit*] the poem strives to dedicate to all who encounter it, its sharper sense for detail, for outline, for structure, for color, but also for "winks" and "hints"—all that is, I believe, no achievement of the daily more perfected apparatuses of competitive (or complicit) eyes. Far rather is it a concentration that remains mindful of all our dates.
>
> 'Attention'—if you will permit me here to cite a word from Malebranche as given by Walter Benjamin in his essay on Kafka—'Attention is the natural prayer of the soul.'

There is an art of artifice. Politicians are masters of that art, and among its acknowledged authorities. But then there is the artless art of the purest poetry, which is also the art of the most natural prayer—an art that has no masters, and acknowledges no authorities.

5.

Que faire? is the title of a book that contemporary French philosopher Jean-Luc Nancy recently published (Nancy 2016), a title that might best be rendered in English as *What Is To Be Done?* At any rate, in either language it is a title with more than one root. In common usage, expressions such as "What's to be done?" or "What can one do?" are typically used, not to ask a question at all, but rather to express resignation. Commonly, they are used as a verbal way of throwing up one's hands in the face of a situation where one's own action no longer seems to have any opening—that is, as a rhetorical way of saying that there really is nothing that one *can* do to change that situation.

As Nancy uses it for the title of his book, that common usage is certainly still meant to be heard. That is one of the roots for his choice of that title. However, the taproot for his choice is not that dismissive common usage. The second part of Nancy's book contains the text of a lecture of the same name—"*Que faire*?—that he first gave in 2012. Early in that lecture, Nancy himself notes explicitly that the main root of his title is one that runs back through Lenin's important text of the same name, a pamphlet that Lenin wrote in 1901 and first published in 1902. Lenin, in turn, took that same title from a famous earlier Russian novel by the 19th century revolutionary Nikolai Chernyshevsky.

Already for Chernyshevsky, then even more emphatically for Lenin, the interrogative sentence "What is to be done?" is not really being used to ask a question at all, any more than it is when used in the ordinary, dismissive way. Neither the common usage of that interrogative sentence nor the usage Chernyshevsky and Lenin make of it really uses it genuinely to ask any question at all, despite the expression taking the sentential form common for asking a question. It is not that one of those two usages of that interrogative form uses it to ask a question, whereas the other one uses it purely rhetorically. In *both* cases, the question form is being used rhetorically—not to ask a question at all, but rather to express something else, something other than a true question. The difference is that in common usage that question-form is used as an expression of *resignation* before the *impossibility of meaningful action*, whereas in Chernyshevsky's and Lenin's usage it is used as a *call to* such action, *revolutionary action* in their case.

In effect, in both cases the question form is being used to communicate an already formulated *answer* to the very question that would correspond to the question-form at issue: the *question* of what is *to be* done—that is, of just what one *should* do—in a given situation. The common usage of the question-form at issue uses it to affirm the answer that there is nothing to be done in the given situation and that, accordingly, what one should do is just that: nothing. On the other hand, Chernyshevsky and Lenin use it to affirm, not that one should do nothing, but rather that one should engage with others in revolutionary activity. In both cases, the question-form is really being used for affirming already arrived at *answers* to the question apparently in question, rather than being used to *ask* that question in the first place.

If, in contrast, one is really *asking* the *question* of what is to be done, of what one should do, then just what is the question one is asking? Even before we ask that question, let alone before we try to answer it one way or another, don't we need to understand just what it is that the question is asking? Don't we in that sense (and not, I warn, in some other, perhaps more common sense of the same expression) need to call the question itself into question, as it were? Don't we, that is, need first to ask ourselves just what the question itself entails, *as* a question?

That, at any rate, is what Nancy does in his 2012 lecture and later (2016) book to both of which he has given that title: "*Oue faire?*" In the positive and non-rhetorical sense I have just tried to indicate, he calls the question "What is to be done?" into question. He does so by examining just what that question itself presupposes or entails. That is, he looks as the *context* within which it would make sense to ask that question—the *world* in which that question makes sense.

When one so calls the question itself into question—not closing it off with an answer, but instead opening it back up and looking at how it is rooted in a context, a world, that allows it to emerge and make sense as a question in the first place—the question of what is to be done shows itself to be one that already takes the matter of "doing" either as a matter of what Aristotle and other ancient Greek philosophers called *praxis,* or as a matter of what they called *poiesis*. That is, the "doing" at issue in the question of what's to be done is always taken in advance by that very question to be a matter of either a *practical* or a *productive activity*.

A *praxis*, to use the Greek word from which the English words *practice* and *practical* derive, is a way of *acting*, as such. If "doing" is understood in that way, then the question becomes one of "politics" and "ethics" as Aristotle defines those enterprises in the first chapter of the first book of his *Nichomachean Ethics* (1094a), namely as "sciences" that aim at "the good" for human beings insofar as that is a matter of "action and choice" (Aristotle 1962). Accordingly, in general the answer to the question of what is to be done is that one should cultivate "virtue" as Aristotle defines it in Book Two of the *Ethics* (1106a), which is as the disposition to choose rightly and well in whatever specific situation one finds oneself in.

Poiesis, on the other hand, is Greek for "making," in the sense of *producing* or *composing* something—for example, shoes, buildings, statues, plays, or poems (a words that itself derives from the same root as *poiesis*). Here, the general answer to the question of what is to be done is that one should do whatever is best designed to produce or compose whatever one is trying to produce or compose. That is, one should employ the appropriate "technology" in the basic sense of that term: the correct "technical knowledge"—from the Greek *techne*, "art, skill, craft," namely of making as producing or composing—for bringing forth whatever it is one aims at bringing forth: a good product or composition of whatever type at issue.

At any rate, whether asked and answered as a *practical* or as a *technical* question, the question of what is to be done is one that makes sense only within the general horizon, as Nancy observes explicitly at one point (p. 97), of aiming at accomplishing some "project, objective, [or] effect. And," as he goes on to remark, "it is indeed understood that all our days we operate under such registers," those very registers being "those that should govern [. . .] prudence," whenever it comes to matters of "regulation, negotiation, strategy," or the like."

However, throughout his 2012 lecture on "What Is To Be Done?" Nancy is concerned with setting apart and confronting an even deeper, very different (though still related, in various complex ways) sense of *faire*, which means a sort of "doing" or "acting," or "making" that no longer has any necessary connection to producing, fabricating, or composing. Thus, after granting the centrality of practical and technical "doing," "acting," or "making" ("*faire*") in all the projects with which we prudently occupy ourselves in our everyday lives, as I have just cited, he immediately adds that neither the practical nor the technical senses of "doing," "acting," or "making," reaches that still other sense that, as he has already noted more than once by that point, is in play in such expressions as—to use just one of his examples, one in which the French "*faire*" needs to be rendered by the English "make"—the expression "make love." (For instance, "Make love, not war," was a famous slogan of the peace-movement in the United States during this country's war in Vietnam.)

To "make love" is not to strive to produce or bring forth some product or offspring. If one fixes one's focus on doing such a thing, as a matter of fact—focuses on "making a baby," let's say—one may soon find one's very capacity for making love vanishing, as it were. Even if one engages in one's love-making in the first place for no other reason than making a baby, one has to forget that focus as it were, if one is going to make the very love one needs to make first, in order to make the baby. Thus here, in making love, the "making" at issue is no producing, even if the ulterior motive for the love-making itself is a productive sort of making. (But then the question becomes whether one is acting and choosing rightly and well in choosing to make love *just* for the sake of producing offspring, rather than just for love-making's own sake. Given that it takes two to make love, one might ask if such purely product-driven love-making is compatible with Kant's "categorical imperative," in accordance with which, in one of its three formulations, one should never treat another person as a means only, and always as an end in oneself.)

At the same time that it is not a matter of making as producing, the "making" at issue in making love is also not a matter of engaging in a certain "practice." Love-making is not in that sense a "practical activity." That is, it is not a *praxis*. I will risk putting the point this way: The only way to make love rightly and well is simply to make it, without worrying about how well and rightly one is doing it.

As Nancy puts it, the sense of "making" at issue in making love is one in which our "making"—our "doing" or "acting," our "fairing" (our "*faire*-ing," we could say, to make a language-mix)—becomes indistinguishable from our living or existing as such. It is a "doing" or "making" that is no longer confined within the horizon of any project or objective, and has no intention of making any particular effect. Instead, making love is just "a part of life," as we might say, a fragment of life into which life pours all of itself without remainder—a fragment that contains the whole.

In a passage I have cited above, Nancy grants that the prudential registers of praxis and production that are presupposed by the question of what is to be done occupy us constantly in all of the projects of our everyday lives. He ends that same passage, however, by noting that the sort of "doing" or "acting" at issue in talk, for example, of "making love" is one which goes beyond those limits. The sense of "doing," "acting," or "making" at issue in such locutions as "making love" is, rather, one in which "to do or to act [*faire ou agir*] is to let oneself be transported to the limits of those registers, to where the impossibility *d'en finir* opens up into the necessity *d'infinir*—if I may risk that term."

I have left part of Nancy's line in the French, because it is not possible fully to capture in English what he is saying in the same way he tries to capture it in French, by the "risk" he takes in his way of wording the point. "*D'en finir*" means "to finish with it," or, to put it differently, "to bring it to an end." French for "end," in the sense of limit or boundary, is *fin* (as in *fin d'siècle,* "end of the century"). The same Latin root from which the French *fin* comes, also feeds into the English *finish.* With all that in mind, we might try to render *"d'en finir"* in the context of Nancy's remark as "to finitize it." If so, then perhaps we could render what he says about the sense of "make" in expressions such as "make love" this way: in such usages, "to do," "to act," "to make," is to let oneself be transported—as one is in making love, for instance—beyond all the projects and objectives and effects with which one is ordinarily occupied, "to where the impossibility of finitizing [what one is doing or making] opens up into the necessity of infinitizing." Or, making yet another effort to capture the in-effectivity, in-operativity, and un-projectability of such making as is involved in making love, to paraphrase: The "making" at issue in such making as that of making love is a making that has no end at all, and therefore opens upon the unending.

6.

Carl Schmitt was a major 20th century German legal and political theorist who provided Hitler's Third Reich—and other regimes similarly founded on the claim to special executive privilege based on the supposed need to address situations of proclaimed emergency—with theoretical justification in advance, in such works as *The Concept of the Political*, first published in 1932 (Schmitt 1996). His thought continues to be influential to this day.

Schmitt maintained that the basic *political* concept is the distinction between *friend* and *enemy*. That distinction, according to Schmitt, is "basic" in the strong and literal sense that it both founds and sustains politics as such, giving politics both its starting point and its continuing support. Thus, the distinction between friend and enemy—between "us" and "them"— is "founding" of politics, by Schmitt's analysis, in the double sense of the Greek *arkhe*, which means both "beginning," as it still does in the English derivatives *archaic* or *archaeology*, and "ruler," as in continues to mean in such English derivatives as *monarchy* or *anarchy*.

Only by first literally *projecting*—from Latin *pro-*, "forth, before, in front of" + *iacare,* "to throw"—that is, by first casting or throwing before or in front of itself what is to count as *other* than itself, can the political community, as Schmitt theorizes it, constitute *itself* in the first place, as a distinct community. It is only by sharply closing itself off from and thereby *re-*jecting whomever it has so *pro-*jected as its definitive "other," that the political community can en-close itself within its own borders, and thereby first clearly define "itself" as a community.

So conceived, as the projective opposition of an "other" or "enemy" to "itself" and its "friends," the *political* as such is in essence *polemical*—from the Greek *polemos*, "strife" or "war." The political society, in Schmitt's conception, is thus the society always at war with one enemy or another, because it can only define itself in term of such enmity. It is a society that can feel secure only in such ongoing warfare, regardless of whether that warfare breaks out into "active hostilities," as they are called, or just remains at what could appropriately be called the level of "passive aggression." The "enemy" must, as it were, be more original, more truly archaic, even than the community "itself" and all its "friends," if the community is to *found* itself by setting itself in opposition to that original and originating enemy.

Conceived along Schmitt's lines, politics is always driven to totalize. Its goal is to define and circumscribe some clearly identifiable, entire, homogenous community, to enclose it, and then to protect it from all intrusions by "outsiders"—all those through the pro-, e-, and re-jection of whom the totalities of such a political community first constitutes, produces, or composes itself.

The masters of politics so conceived, those who can lay claim to political authority and make good on that claim, are those who most effectively manipulate the polemical mechanisms for *dividing* "us" from "them," and then keeping the former stirred up against the latter. With regard to war thought of as "active hostilities," we might invert an old line from von Clausewitz and say that politics so conceived is just the continuation of war by other means. Accordingly, successful politicians would be those who proved proficient in fomenting and fanning enmity. The art of dividing the community, then keeping the division going strong, would be the art in which they were most gifted. The art of politics so conceived would be the art of using fragmentation to produce and maintain a potentially endless number of mutually exclusive human wholes—a demagogic art, designed to produce far more heat than light, of course, until finally all the light is blotted out.

However, borrowing from Celan's reflections in "The Meridian" on the distinction between the art of mere artifice and the artless art one finds, for example, in what he calls the "pure poem," we can perhaps distinguish such arty and artificial politics—the politics of grandiloquence, pomp, demagoguery, and coercive power in general—from a "pure" politics that would no longer have anything of such artificiality. Such pure politics would be, instead, utterly "natural." As pure, natural politics would thus be more original than any politics such as Schmitt conceives of politics, more truly foundational than any conceivable opposition between "friend" and "enemy," "us" and "them."

Pure, natural politics would then be a politics that truly built up the *polis*, the "city," as the very place of genuinely and fully *human* habitation. It would be the politics of building and maintaining that place, the polis, as a place where the distinction between friend and enemy, "us" and "them," would no longer have any place. That polis would have no need any longer to enclose itself within walls, since it would open itself to all, each and every one. It would be a political community that would need to practice war no more, since there would be no one there any longer to go to war against. Such a pure and natural polis would be precisely the place where "politics as usual," that very politics that is no more than war by another name, would at last become *inoperable*.

Only such a pure, natural, and original politics of the inoperability of "politics as usual" could join up with the pure, natural, and original aesthetics that would be up to thinking such pure art as Celan's gives us to think. And only the conjunction of the three—pure art, pure politics, and pure thought—would give us any real hope of saving the fragments of nature that will be all that remains when all the wars, including all the wars against nature herself, are over.

That is so, regardless of whether wars' end itself comes in time—or only when we are all out of it.

7.

Finitude is the opening to the infinite.

At one point in the text from his 1941–1942 seminar exercise in philosophical thinking, Heidegger contrasts two different senses of "un-alive." First, he uses the example of a piece of chalk. If we say that the chalk is "un-alive," we simply mean that it is not a living being at all. The chalk is, instead, what in English is called "inanimate," a word that itself derives from the Latin for "un-alive" or "non-living" (from *in-* as a negative prefix + *animatus,* "endowed with life"). On the other hand, if, for example, a teacher says of the students in some class on a given day that were very "un-alive," the teacher is not saying that the students were inanimate objects. At least that is so, unless that expression "inanimate" itself is being used differently than when we say that a piece of chalk is such an object, so that it means they "may as well have been pieces of chalk" or the like.

The teacher might also have conveyed the same message by remarking that the students were very "dead" that day. That is itself not something we would say of an inanimate object such as a piece of chalk—that it was "dead"—at least in the same sense that we would say it of such students. Once again, we might say something such as that an especially unresponsive group of students was "*as* dead *as* a piece of chalk," but that is just a way of saying something along the same lines as "they might as well have been chalk, for all the responsiveness they showed."

Later in the text of his 1941–42 seminar exercise in philosophical thinking, in one of the protocols for one of the class-sessions, Heidegger expands upon the point and in the process relates his discussion to the distinction between friends and enemies (the very distinction discussed in the preceding section of this very essay, the one foundational for "politics" as Carl Schmitt, Heidegger's contemporary, conceived it):

> 'Un'- is a negation, corresponding to the Latin *in-*, to the Greek *a-* or *an-*. [. . .] What is not alive, un-alive? All that is either dead or lifeless. By 'lifeless' we mean the non-living that is certainly real, but to the reality-character of which life does not belong. [On the other hand,] only that which is *no longer* alive can be dead. A frozen bird is un-alive, that is, dead. The chalk is un-alive, that is, lifeless. The audience is un-alive. This time, 'un-alive' is used in the sense of unawake, not lively. These examples are meant to show that with the 'un'-, as negation alone, nothing is said yet. What counts is what is negated, and how what's negated is fixed in the negation. The first sort of negation [in the example of the frozen bird]—such un-aliveness, such no-longer-living—presupposes life. This negation is not a mere nothing, but a [. . .] privation, an exclusion. [. . .] We rely precisely on that which we exclude. That pertains to the essence of every exclusion, as also to the essence of all enmity. In enmity there is always a dependence on one's enemy, that against which one defines oneself.

Thus, what is negated in "privative" or "exclusive" negations is always *retained* in its very negation, insofar as the identity established by such exclusion is precisely a function of what is thereby excluded. The lack of life, of animation, in an unresponsive audience, for example, remains itself, as Heidegger notes, "a determinate kind" of life, of being alive—a determinate "kind" or "sort" of life: the German term at issue being "*Art*." However, as Heidegger is careful to note: "Here, the concept of kind [*Art*] is used not in the sense of a sub-order of a class [in effect, as a species of a "genus," which a common translation of the German term at issue: *Gattung*] but in the sense of sort and manner [*Art und Weise*]."

It is no less worthy of note that the sense of "kind" at issue here is the same as that in which we speak, for instance, of "kinds" of art or, for that matter, of "kinds" of people, and do not mean thereby to suggest any idea that either art or humanity is some sort of general (literally: "of or pertaining to a genus") class that has various sub-classes (or sub-types: "species"). The various "kinds" of art, whether one means thereby such different kinds of art as music, painting, sculpture, or dance, or rather such different kinds of art as Impressionism, Expressionism, Conceptualism, Modernism, or Post-modernism, are *ways or manners in which art can be*, not species of some genus, sub-classes of some class.

There is not some one, common feature that belongs to all the things we call "art," such that they all belong to some one large class or genus of things, and which can then be divided into sub-classes or species by virtue of other features that some but not all of the members of the general class at issue share. In the same way, Africans, Americans, Europeans, and Asians are not "species" of the genus "homo." Humanity is not some natural kind, nor are different kinds of people different species of some one genus—regardless of what racists would have us believe, for their own impure "political" purposes.

Nor, for that matter, is making love a species of some genus of doing or acting or making. Making love, and not war, is not a matter of producing one species of offspring rather than another, or making one sort of product instead of another. Instead, the sort or kind of "making" at issue in such things as making love, making war, or even "making nice," is a way or manner of making that escapes all the categories pertaining either to procreation or production or even praxis, and becomes no more and no less than a way or manner of being, of living, of human existence itself, as Jean-Luc Nancy teaches in *Que faire?*

Nancy calls to our attention, as noted above, that it is precisely because the kind of making at issue in making love or the like does not aim at any offspring, creation, or effective action, that it is without end. Such making, such doing, such *faire*, is thus "without end," or "endless." It is endless, however, not in the sense that it just goes on and on forever, but rather in the sense that it has no "purpose" or "goal" at which it aims. It is just itself. It is not "for" anything, any more than chipmunks are. As I used to like to say to my own classes, all that chipmunks are "for" is chipmunking.

Aristotle said the same thing in a more general way long ago, when he argued that the end or purpose of any thing of a given "natural kind," such as a tree or wolf or any other thing that came about by "nature" (*physis*) rather than by "art" (*techne*) was just to be the very kind of natural thing it was. "The purpose of cow is cow," as one of my own old professors used to like to put it.

Many centuries after Aristotle, the poet Gerard Manley Hopkins also made the same point poetically, in the first stanza of one of his poems, at least by one way of reading it (Hopkins 2013):

> As kingfishers catch fire, dragonflies draw flame;
> As tumbled over rim in roundy wells
> Stones ring; like each tucked string tells, each hung bell's
> Bow swung finds tongue to fling out broad its name;

> Each mortal thing does one thing and the same:
> Deals out that being indoors each one dwells;
> Selves—goes itself; myself it speaks and spells,
> Crying *What I do is me: for that I came.*

8.

"The aesthetics of nature" is an ambiguous phrase. Its meaning flows in two directions at once, depending on how one takes the preposition, the "of." Taken one way, the aesthetics of nature would pertain to how we address nature. Taken the other way, it would pertain to how nature addresses us.

Henri Nouwen (1932–1996) was a Roman Catholic priest. He was raised within that same tradition of Catholic, Latin-based, "Western" Christianity. Nevertheless, he was always open to learning from traditions other than his own, toward none of which did he harbor any enmity. In *Behold the Beauty of the Lord: Praying with Icons,* Nouwen explores the Orthodox, Greek-based, "Eastern" Christian practice of doing what the subtitle of his book says: praying with icons (Nouwen 1987).

With regard to his own Latin Christian tradition, which has historically tended to emphasize *verbal* prayer, Nouwen points out that even within that tradition praying does not require the one who is praying to do all the talking, as it were. Rather, believers also—in fact, most especially—pray when they stop talking and just *listen*, allowing the silence itself to find voice.

In Greek Christian tradition, emphasis on the verbal dimension or prayer has not been so exclusive. There, the visual dimension of prayer receives greater attention, above all in the practice of praying with icons. There, the emphasis in prayer has been less exclusively focused on *hearing* and has instead made more room for *seeing*. It has focused less exclusively on *listening* and left more room for *gazing* as well.

In both Latin and Greek Christian monastic tradition, it became hallowed practice to take short verses from Scripture, memorize them, then repeat them over and over to oneself, literally "ruminating" upon them, like a cow chewing its cud, until it has finally extracted all the juices from what it keeps chewing over. Such repetitive, ruminative, as it were "bovine" practice was itself regarded as *already* being prayer. And only after long practice of such praying, such chewing over and over again of the morsel of holy text, would that text itself finally reveal its full power to the one praying it—such revelations sometimes coming with sudden force, like bolts of lightning, at other times more gently opening themselves up to the practitioner of such prayer.

Similarly, as Nouwen interprets the practice of praying with icons, the meaning and power of an icon does not typically reveal itself at first glance, or even after many glances. Typically, it is only after prolonged repetitive practice of looking prayerfully at the icon—visually memorizing it, in effect, then repeating what one had memorized again and again and again, over sometimes very prolonged periods: thereby *praying* with the icon—that one begins truly to *see* it.

Practicing such seeing is actually a practice of looking not so much *at* what one sees as *through* it, to what it in turn gives to be seen. That is even and especially so with regard to what can never be seen directly, but always only seen *through* what can be directly seen.

Listening, diligently and regularly practiced, eventually on some rare occasions may finally let one hear the silence itself that is broken by all the words being spoken, broken to give that very silence, itself voiceless, a voice. Just so does gazing, if practiced long and devotedly enough, sometimes, on rare occasions, finally let one see the invisible, making visible the invisible itself.

When that happens, one is tempted to stop gazing—a temptation always to be resisted, if one is not to lose sight once again of what can never be seen, which is what one is always most given *to* see. When one yields to that temptation, one falls into idolatry. That is, one turns the icon into an idol. An idol is that which fixes the gaze on itself, rather than sending the gaze beyond itself, open to what lies beyond every gaze, and what thereby sustains all gazing.

However, if one succeeds in resisting such temptation, which means if one keeps on praying with the icon, keeps on practicing such prayer with it, then eventually, on even rarer occasions, something further may happen. One may realize—whether suddenly or gradually, whether consciously or unconsciously, does not matter, finally—that one is not so much seer as seen: that in praying with icons one is being seen oneself, even more than one is doing the seeing. In short, one sees that "seeing icons" is an ambiguous phrase, one that flows in two different directions of meaning at once, depending on who is seeing whom. In gazing long and well enough—which is to say prayerfully—at icons, we come to see that they have all along been gazing at us. Or, rather, we come to see that what lies beyond all gazing has been gazing at us through the icon, as through spectacles.

Early on in his book on praying with icons, Nouwen writes that "it is only since I started to view them [icons] as the iconographer intended—not as decorations but as holy places—that they have told me their secrets" (Nouwen 1987, 13) Finally, the secret of an icon is just this, that if we practice looking at—and through—an icon long enough, we will find in the end that the holy is looking at us through it.

"By facing us," writes Bruce Foltz at one point in his recent book *The Noetics of Nature: Environmental Philosophy and the Holy Beauty of the Visible*, "the icon not only engages us, but addresses us" (Foltz 2013, 140). Following Nouwen, we might put the same point even more emphatically: By facing us, the icon not only gives itself up to us to see, but also gives us up to *be* seen.

An icon is something that makes a spectacle of itself in order eventually to make a spectacle of us in turn—after, by long practice, we finally learn to let it.

9.

"The beauty and holiness of nature," Bruce Foltz writes in *The Noetics of Nature* (just a few lines before his already cited remark that "the icon not only engages, but addresses us"), "do not offer themselves to the mere spectator, to the theoretical gaze." Along those very lines, one might well ask if a spectator who is such a "mere" spectator—which means one who gazes with no more than what Foltz thinks of as a purely "theoretical" eye, an eye that always seeks to *totalize*—ever really sees *anything*. One might well wonder whether the eye of such pure "theory" divorced from all "practice" ever really sees any single, solitary thing *in* the whole—which is to say anything that truly *is*—as just what it is. One might wonder if the eye of a supposedly pure "theory" (from Greek *theorein*, "to look at") that excludes all "practice" has not by such exclusion already rendered itself sightless, unable to see anything at all in its own simple being in the whole, which means precisely see it in its own "beauty and holiness." In its rush to totalize its vision, such a purely "theoretical gaze" literally over-looks the only beings that are ever given at all, which are always and only given as beings *in* the whole, not as parts only subsequently subtracted and abstracted from some whole of beings that can only be *projected* as given, without ever truly *being* so, to *any* gaze.

A few pages earlier, Foltz has already written of the "nature" revealed only to the gaze that is no longer purely "theoretical" gaze but is, rather, open to the "beauty and holiness" of what it sees, that such non-theoretical nature "is divine gift, self-expression, and image. It is iconic" (Folz 2013, 130).

At the very beginning of the modern epoch, that very epoch that has led today to such global environmental destruction that earth as a whole is threatened by it, Galileo famously proclaimed mathematics to be the language of nature. That is indeed the language of that "nature" to which modern science everywhere and always listens, a "nature" that excludes whatever cannot be expressed mathematically. There is nothing iconic about such a "nature"—nor anything *poetic*.

There is, however, an older, other "nature"—a more original origin, a more natural nature, I will venture to say, than that of modern "natural" science. That more natural nature *is* iconic. It does not give itself to the purely "theoretical gaze" Foltz mentions. Instead, iconic nature gives itself only to the person who views it with an eye practiced in seeing concretely. It reveals itself as iconic, as Foltz also remarks, "only to the person who approaches it with piety and in the belief that it is holy" (Foltz 2013, 138).

When such natural nature shows itself, it shows itself in and as an icon. By the same token, when it speaks, it does not speak the language of mathematics, that language for which alone modern science has ears. It speaks, instead, the language of poetry, of the pure poem.

Physis is a Greek word that is ordinarily translated as "nature," and from it derives our English name for the science of "physics," the science that supposedly investigates "nature." However, the *physis* of physics, most especially and pointedly of modern physics since Galileo, is not the *physis* of poetry: The nature investigated by physicists in their science of physics is not of the same nature as the one sung by poets in their art of poetry. That is true above and before all of any poetry that is truly "after the Holocaust," as Celan's "pure" poetry is.

The language of poetry is not the language of science, nor is it one and the same nature that speaks both languages. The nature that speaks the language of poetry is much dirtier, much "earthier," than the nature that speaks the language of mathematics. Correlatively, the former language, that of poetry, is much *too* "earthy" for ears that can only want to hear the nature that speaks mathematics (that never-ending chatterbox of a language that, as Gödel knew, can never be completed).

In his lecture "On the Origin of the Work of Art," the first version of which he delivered in 1935, Heidegger ventures to translate the Greek term *physis*—which is ordinarily translated as "nature" ("*Natur*," in German) and from which we get the word "physics"—as "earth" ("*Erde*"). That, too, is a matter of the ears with which one hears what the Greek word says, when taken in the full concreteness of its context in that now-global "Western" history that goes back to the ancient Greeks—and to such ancient Greek philosophers as Thales, Anaximander, and Anaximenes, then Heraclitus and Parmenides, followed eventually by Plato and Aristotle, and so on, down to Hegel and Nietzsche and various hangers-on after them.

In "On the Origin of the Work of Art" Heidegger says that the origin, the wellspring, of the work of art is the strife or conflict (*Streit* in German) between such an "earth" and what Heidegger calls "world" (Heidegger 2002). The work of art is the place where the contest between the two, "earth" and "world," is contested. It is the place where the fight between the two is fought out.

Already in *Being and Time*, first published in German in 1927, Heidegger articulated the concept of "world" as being that of the whole "wherein" every being is. But, as I've already emphasized before, that "whole," that "world" in Heidegger's sense, is no sort of Wittgensteinian totality of everything that is the case—that totality which, as Kant already taught, is *never* given. The world, the whole "wherein" whatever is *is*, is no *Ganze des Seienden*, no whole of beings, to use the terms Heidegger does in his inaugural lecture in Freiburg in 1929 ("What Is Metaphysics?"), two years after *Being and Time* first appeared. Rather the "world," as Heidegger articulates it *Being and Time* itself (Heidegger 1962, H65), is "that '*wherein*' [we] can be said to 'live'," and in the context of which we encounter whatever beings we do encounter—whatever beings are given *in* such a non-totalizable whole.

The connection between "earth" and "world," their interrelation one with the other, is a *contestation* or *strife*—that "war," *polemos* in Greek, which, according to fragment 44 from Heraclitus (Heraclitus 1935, Bywater numbering), is "the father and king of all things." Yet the war at issue is no Schmittian battle between "friends" on one side, and "enemies" on the other, no war between "us" and "them." Rather, it is a battle or contest between friends, the sort of struggle that calls out the best in both, as true friends always contest with one another to call forth the best in each other. Such a struggle, contestation, or strife—such a *polemical* interrelationship—is always in that way a genuinely *loving* one. Such warfare is an *embrace*, and not any sort of compulsively projecting-rejecting exclusion. That is exactly why such warfare is generative, as Heraclitus said, rather than destructive, as are all wars between "us" and "them."

Modern "natural" science, the paradigm of which is modern physics, is based upon the prior projection of the "nature" it investigates as the totality of a mathematically determinable manifold. Alain Badiou, for whom mathematics itself is the science of pure being itself, as sheer multiplicity, sees that. So did Descartes already see it, as his analysis of the piece of wax in the second of his *Meditations* demonstrates.

Such a *physis*, the "nature" that modern physics investigates, takes no part in the generative, loving war of "earth," that very different *physis*, with "world," what the Greeks called *kosmos*.

The modern science of nature, of *physis* conceived as a mathematically determinable manifold, can never enlist with nature, with *physis*, cast in art as earth, to engage with world, with *kosmos*, in contestation. That is because, as Heidegger saw and repeatedly said, the projection of being—that which *is*—upon which such science is based is one that always and everywhere *de-natures* nature, and *de-worlds* the world, letting neither *be* in unending strife with the other.

What *science* leaves behind once its work is over is no longer "world" at all. It is sheer wasteland, where nothing grows—that wasteland that, as Nietzsche said, itself just keeps growing. Only wars of devastation can be fought on such a blasted battlefield. It is no place for the loving war between earth and world.

In contrast, *art* is a battlefield on which that loving war between earth and world—or "nature" and "world": *physis* and *kosmos*—*can* be, and *is*, fought out. It is one way in which what we call nature can be cultivated, rather than dominated. As is always true in such cultivation, what we cultivate cultivates us in turn. It is only in the cultivation of another that we become most ourselves. By cultivating nature, we let it cultivate us: When we see to it, it sees to us.

That is the aesthetics of nature.

BIBLIOGRAPHY

Aristotle. 1962. *Nichomachean Ethics.* Translated by Martin Ostwald. New York: Bobbs-Merrill.

Badiou, Alain. 2005. *Being and Event.* Translated by Oliver Feltham. New York: Continuum.

Badiou, Alain. 2009. *Logics of Worlds: Being and Event, Volume 2.* Translated by Alberto Toscano. New York: Continuum.

Celan, Paul. 1961. *Der Meridian.* Frankfurt: S. Fischer.

Foltz, Bruce. 2013. *The Noetics of Nature: Environmental Philosophy and the Holy Beauty of the Visible.* New York: Fordham University Press.

Heidegger, Martin. 2008. "*Einübung in das Philosophische Denken.*" In *Seminare (Übungen) 1937/38 und 1941/42* (*Gesamtausgabe* Band 88). Frankfurt: Vittorio Klostermann.

Heidegger, Martin. 2002. "On the Origin of the Work of Art." Translated by Julian Young and Kenneth Hayes. In *Off the Beaten Path.* Cambridge, UK: Cambridge University Press.

Heidegger, Martin. 1998. "What Is Metaphysics?" Translated by David Farrell Krell.In *Pathmarks.* Cambridge, UK: Cambridge University Press.

Heidegger, Martin. 1962. *Being and Time.* Translated by John Macquarrie and Edward Robinson. New York: Harper & Row.

Heraclitus. 1935. Surviving fragments. Translated by Richard Lattmore. In Matthew Thompson McClure, *The Early Philosophers of Greece.* New York: Appleton- Century-Crofts.

Hopkins, Gerard Manley. 2013. *Complete Poetical Works.* London: Delphi Classics.

Milner, Jean-Claude. 2014. *La puissance du detail: phrases célèbres et fragments en philosophy.* Paris: Éditions Grasset & Fasquelle.

Nancy, Jean-Luc. 2016. *Que faire?* Paris: Éditions Galilée.

Nouwen, Henri. 1987. *Behold the Beauty of the Lord: Praying with Icons.* Notre Dame,IN: Ave Maria Press.

Rilke, Rainer Maria. 1995. "Archaic Torso of Apollo." Translated by Stephen Mitchell.In *Ahead of All Parting: Selected Poetry and Prose of Rainer Maria Rilke.* New York: Modern Library.

Schmitt, Carl. 1996. *The Concept of the Political.* Translated by Geroge Schwab.Chicago: University of Chicago Press.

Wittgenstein, Ludwig. 1922. *Tractatus Logico-Philosophicus.* Bi-lingual edition.English translation by C. K. Ogden. London: Routledge & Kegan Paul.

NOTE

1. All translations from sources listed by French or German titles alone in the bibliography at the end of this essay are my own.

Chapter Two

The Noetics of Poverty in the Paintings of Vincent Van Gogh

Joshua Coleman and J. Porter

> The world is one . . . for the *spiritual world* in its totality is manifested in the totality of the perceptible world, mystically expressed in symbolic pictures for those who have eyes to see. And the *perceptible world* in its entirety, when it has been simplified and amalgamated by means of the spiritual realities. The former is embodied in the latter through the realities, the latter in the former through the symbols. The operation of the two is one. —Maximus the Confessor, *The Mystagogia*

> From a Deep Dream I woke and swear
> The World is deep,
> deeper than day had been aware.
> —Friedrich Nietzsche, *"The Other Dancing Song,"* Thus Spoke Zarathustra

PART I

As early as the seventh century B.C., evidenced in *The Iliad* and *The Odyssey,* the Greeks made a critical differentiation between natural and human phenomena. Long before Socrates it was customary to distinguish between those things considered static—plants, animals, the stars and so forth—and those in flux which were consecrated by human beings such as customs, laws, political institutions, etc.[1] The former became known as the *physis,* loosely translated as nature, and the latter as *nomos* or law—more loosely, culture or custom.[2]

This distinction becomes more deeply ingrained after Aristotle, who separated what comes about by nature (physis) from what comes about by art (techne), implying that the latter is a product of human culture which has little to do with the presumably deterministic, mechanistically independent *physis*.

In contrast, 6th century Greek monk, Maximus the Confessor, suggested that in nature itself we find "the clothes of Christ" illumined at the Transfiguration on Mt. Tabor. For Maximus, however, it is not that Christ and nature suddenly "lit up," as it were. Rather, he posited that a change occurred in the perception of those allowed to *noetically* perceive Christ and nature as they always shone. This would suggest that nature is not static, much less determined or mechanistically independent, as had been assumed in the inherited cultural distinction.[3] Maximus writes about those privy to this event, "They passed over from flesh to spirit before they had put aside this fleshly life, by the change in their powers of sense that the Spirit worked in them, lifting the veils of the passions from the intellectual activity that was in them."[4] What is first "witnessed" here is the change in themselves, one which allows them to perceive the real in terms of depth, an apprehension of being beneath the immediate empirical, such apprehension previously unavailable even while Christ and nature had been right in front of them all along.

The implications are far reaching, as the healing of this culturally constructed divide between culture and nature is critical to a theological aesthetics that facilitates the transformation of relations between persons and nature. This is in distinction from an approach which views nature as merely a "moral responsibility"—yet remaining essentially separate or "other"—and which by itself is unable to maintain relevance against contemporary and exploitative cultural institutions.

The role of *techne* or art is also central to the Eastern Christian context of Maximus's interpretation, specifically the role of iconography as a sort of impressionism that calls on the imagination of the viewer to participate in the scene loosely depicted; and in so doing, change his perception. In this sense, *techne* reveals the bridge between *nomos* (culture) and *physis* (nature), a deep inseparability already there but not perceived within certain postmodern, materialist presuppositions toward and about nature. Unlike for Aristotle, we will see that *techne* or art can reveal the inherent communion between culture and nature, and that art's role in this process always has been (and continues to be) critical in order to engage the natural world as an authentic extension of our own being and place of dwelling.

Applying an understanding of symbol in its traditional sense as *symbolon,* or that which brings together two halves, specifically an invisible presence through the visible natural world, Maximus writes unequivocally about nature's role, that Christ's clothes are a symbol of "creation itself, which a base presumption regards in a limited way as delivered to the deceiving senses alone, but which can be understood, through the wise variety of various forms that it contains, on the analogy of a garment, to be the worthy power of the generative Word who wears it."[5] Nature, then, is not an objective "other" for a distanced stewardship or moral responsibility, but, as a divine dwelling, is in constant relation to us, our ability to perceive and respond depending upon our own spiritual perception and progress.[6]

Likewise visual art, with its appreciable variance and divergence, is born of a flux between viewer and artwork. This relational fluidity depends in large part upon the participatory capacity of the viewer, even while the artwork remains tied to the material world. It should be noted, however, that to speak in terms of capacity is not to assert a uni-directional notion of progress, tempered by a single moral position, or the facile consolation of a single abiding totality or testimony, but a thickening of the physical world, an expansion beyond the immediate-empirical to further reveal the essence of nature's unsettled disclosure.

These two approaches to *techne,* one as the bridge which reveals the coherence between nature and culture, and the other which focuses on a culturally constructed divide between them, are each found in the early works of Vincent van Gogh. His depictions of the poor in the first part of his career display worn faces and clothes represented by subtle, earthy tones, ones that allow the characters to blend into the surrounding landscape, providing a depth of realism which reveals a mutual influence between nature, man and his sufferings. Often the faces appear as gnarled roots, eyes blackly round and gold like empty birds' nests, limbs heavy or like pollard birches, all of which were subjects of study at the time. In an attempt to assault the beholder, he rendered them in exaggeration, a kind of gaudy innocence, but which was neither unnatural nor caricature.

Depictions of the rich, however, refused such "drab" tones, demanding brighter colors to set apart a *culturally* created beauty in distinction from, or in complete obliviousness to, that of the natural surroundings; in this case the creation by *nomos* or culture is in antagonistic separation from *physis*. As will become evident, our cultural presuppositions about beauty enclose the space of dwelling, typically with nature as background (regardless of distance), and man as foreground, with the modal standards of capitalism partially composing the image before exposure, like a camera, immediately commodifying the natural world.

This need not mean that the use of bright colors is a departure from nature, as in other places Van Gogh applies bright colors within the context of an impressionism that successfully collapses the distance between culture and nature, foreground and background, object and horizon. One such instance is his still life of fish, where he juxtaposes the brightness of the fish with an impressionist backdrop.

In this instance, however, rather than presenting an idealized *techne,* the focus is that the fish are dead, death itself leaping out to the viewer who holds the gaze, perhaps a more piercing death in ironically loud tones. Another possibility is that the emphasis on death through bright colors is a symbol of cultural beauty being the death of nature. Regardless, the use of loud color need not mean an automatic identification with cultural beauty, as will be discussed, but sometimes its restless brightness serves as a vanishing point, one to which the beholder's eye cannot help but repeatedly return; creating the spot wherein the eye finally catches itself going back repeatedly to the place it wants to enter and go through but cannot, catching itself roving in a double personality: part of the immaterial and material world, part mind and part appearance. In this case the brightness serves almost as a rope, a noose dragging the viewer in, made from the entwining fibers of death and culture.[7]

Thus our preconceptions of cultural beauty directly impact our aesthetics of nature. On this matter contemporary thinker Bruce Foltz refers to the work of intellectual historian Marjorie Nichols when he writes:

> . . . until the last few centuries, the Alps were regarded as so repulsive in their bleak, chaotic, barren disorder that coaches traversed them with curtains drawn, to protect travelers from such a repugnant spectacle. The wild nature we work to preserve in the harsh mountain high-country, the scoured canyons of the American Southwest, and the flooded marshlands and estuaries of the American Southeast is beautiful, but it is a beauty that is, in an important sense, a creative product.[8]

A few important points converge here. First, the appreciation of natural beauty coalesces with a cultural lens placed upon it, meaning natural beauty is to some degree attached to culture. Of course, the need to be consciously aware of humans' interactive role in creating and perceiving beauty in nature is accelerating as the world, in general, seems to be adopting a pervasive ideology of pure materialism. Part of Foltz's point is that contemporary Western, Euro-Capitalist culture and its offshoots have fostered a dispersed, abiding, and nearly global consciousness that we possess an earth essentially retentive, in that we must harrow its life, inorganic and organic, not primarily to sate human need or desire, but for profit, by which is thereby ascribed a tentative ontological status on the basis of this gratification alone, rendered meaningless and devoid of spiritual mystery. Indigenous cultures, and at one time the great majority of all cultures outside the Modern West for that matter, saw a certain mystery and awe in nature. Within a cultural inheritance of nature as sacred, certainly the constructed divide between culture and nature is more difficult to arise as an idea, much less take root as something perceived to be essentially real.

Based on the above Foltz quote, the ability to perceive beauty through nature in the Modern West must consciously include a transformed conception of beauty upon removal of reductive, material conceptions that grossly limit its perceptive potential in persons. Obviously, much has changed since the view of the Alps as "repugnant," and in no small part due to nature writing, be they directly descriptive essays or more nuanced poetic imagery.[9] Such writing makes clear the important link between imaginative constructions of aesthetics and those practical applications and general perceptions of nature as worthy of cultural care.[10]

Part of the mystification caused by the inherited division between culture and nature is the notion of nature as an independent, self-regulating mechanism or as an objective other in opposition to culture, the latter the result of human imagination and creative interaction. Foltz writes:

> The concept of nature as autonomous, self-subsistent or self-contained, has for several centuries confounded modern thinking about nature. More recently, it has blinded environmental philosophy both to the unitary character of the problems that it faces, as well as to the direction to which it needs to look for answers. It is a conception of nature, and stands for a specific relation to nature, that is inherently one-sided, for the cosmos that it discloses possesses only one side.[11]

This basic assumption has led to a destructive misuse of language with regard to all things deemed "natural." More often than not, when we use this term, we tend to ascribe a certain destiny and self-determination to whatever is at issue. Something is "natural" when we determine a sort of inherent *fait accomplis* when left to its own devices. Yet if anything has been made clear from the current environmental crisis, it is that nature is not a self-determined "other" in independent separation from us, and, in the case of the natural world, that which is "natural" is in a fluid dependence between human creativity (as both extension and cause) and its own flowering.[12] Moreover, our categories of separation between culture and nature only serve to cloud our ability to perceive our intimate place in and with the natural world.

In fact, a disassociation between nature and beauty has, since as far back as the Greek inception of Western thought, been tied in with *Logos*. It belies how faithbearing we are willing to be toward the unknown, as subset of our deep animosity with earth as the place of pulchritude, while at the same time refusing to link our essential nature to it. Man has always sought beauty. But part of beauty is essentially *inhuman* and *other,* just as *eros,* by definition, seeks what it has not. We love beauty, like wisdom, because of a certain absence in us, and yet we have a capacity and hunger for more than what we have. As Hannah Arendt noted from Pericles' Funeral Oration: "Men love wisdom and therefore begin to philosophize because they are not wise, and they love beauty, and do beauty, as it were . . . because they are not beautiful."[13] The willingness to admit these lacks in ourselves throws open the desire and capacity to recognize and engage their reality as presence in the natural world.

By craving what one admittedly is not and has not, *eros* secures the love of beauty as non-narcissistic, but it also means that the beloved is always and everywhere missing. Beauty is the other, and therefore *in part* missing from the natural world from which man materially derives and shares the lowest piece of his nature; our desire for meaning leads to lack, a desire to love what is not here, or at least not materially immediate. Arendt writes: "Because thought's quest is a kind of desirous love, the objects of thought can only be lovable things— beauty, wisdom, justice and so on."[14] Beauty dwells in the sphere of forms, of *Logos,* of which earth is a dim shadow at best, at worst it dissolves ontologically into a status of original meaninglessness, by its material opposition to the real things of thought.

In this inherited Greek philosophical approach to nature, we attribute a begrudging destiny and self-determination by default, because at the highest level of Being and metaphysics, it is not real, lost to constant change and incapable of the stability of rational truth as Plato conceived. Yet it also remains the humiliating base into which we will physically descend. As the cruel shepherd of our flesh into oblivion, nature, like the eye, has this double life of being real and not real, thus seen but not apprehended, in the taxonomy of the soul.

Given the huge impact of nature writing and art upon our ability to perceive beauty in nature, it is imperative to articulate a philosophical and theological aesthetics of nature that will further influence the way we see into and respond to nature as more than material, bringing that aspect of beauty that is materially absent into perceptible being. Such an articulation could provide a middle ground between the extremes of nature as material and nature as romantic muse for poets. This can also facilitate a departure from the logic that nature is an objective other to "fix," altering our relationship to it without holding it upon an untouchable pedestal which, as will be later discussed, actually furthers inaction with regard to the current dilemma. In expressing nature as a place of dwelling, we find in it a facilitator of a way of being necessary toward becoming fully human. Thus we have a circular hermeneutic; dwelling in nature teaches us how to dwell in nature.[15] Likewise, perceiving beauty in nature teaches us how to perceive beauty in nature.

In terms of perceiving such depths of being, Van Gogh declared that the nude *for the nude's sake* proved lacking in meaning because it replaced a human subject with an idealized *techne;* and that an "ugly" woman (according to the cultural norm) was inherently more beautiful through her ability to express something true about the human condition something that forced the viewer to either actively view in terms of depth, or simply look away. The traditional view of the idealized nude is one whose proportions are essentially the embodiment of *Logos*, following precise mathematical proportions which historically vary in measurement but rarely in ratio, over and against the confluence of biology and experience, the latter of which removes itself from literal representation. Opposed to drawing the nude as a derivative of a sterilized form, Foltz makes the point that nothing is re-presented in iconography. In fact, it is the lack of realism which facilitates a presence presented for the first time; not the copy of a past event, but the facilitation of its present reality. Here Foltz distinguishes this approach as "aesthetico-theology" as opposed to the "onto-theological" tradition.[16]

Once again, this view of aesthetics and this conception of beauty in terms of lived context and depth depends upon the perceptive capacity of the viewer, as in Maximus's view of the Transfiguration. Indeed, in both these latter cases the viewer becomes participant, over and against being a passive recipient of an idealized form. The viewer moves from observer of an object to participatory witness in an event.

The inherited distinction between culture and nature, as critiqued by Van Gogh's work, also suggests a misunderstood relation between beauty, nature and poverty. Splitting art from nature puts the viewer of the former at a distance from the latter, insofar as the imagination of the viewer is no longer called upon to facilitate entry. All too often art is perceive as the object in the gallery, the viewer ingrained in those cities which have cornered off and made rational "natural beauty" through its parks, but in a context where nature is not a place of dwelling to the extent it impacts the viewer of the work, even if the work seeks to convey something of natural beauty. Rather, as a cultural construct, unreal, idealized beauty infiltrates the idea projected onto nature. This allows the viewer's imagination to go dormant and maintain a false, objectifying distance. Van Gogh challenged this false distancing, but in a most peculiar way to unsettle foreground and background. His depictions of farmers bent to the soil, their skin and clothes blending with the earth tones. They *are* the nudes. Naked in the field in the sense that they must keenly feel and intertwine with the vicissitudes of land and air, abandoned in the pretense of fulfillment to the laws of nature so that in Van Gogh's work, especially his early paintings, the peril of beauty is left to the poor. The farmers are depicted more nude than the nudes. For it stands that the studio nude, the model in stasis at the center of academic painting, attenuated of all context, is a way of covering, created by certain kind of ideology; is actually the subject of a certain phenomenological hiddenness. Such objectification creates the capacity for aesthetics to warrant the "seer" as appraiser over any ethical relation, and exonerates her from the political ramifications that ensues such appraisal. Yet the practice of restraint amplifies related cultural issues with regard to the culture/nature divide, and most obviously with the poor.

In terms of *laissez faire* Capitalism, made easy is the dismissal of the impoverished who are most often relegated to the worst environmental conditions.[17] As we come to see nature mechanistically independent, self-determining and out of the purview of our influence, we quickly apply this same logic in a way that sees a created economic system as "natural" by ascribing the same false autonomy to it. It is easy to then wash our hands of its faults, to say that it is "natural" that many will be dispossessed by that system. When combining this view of poverty with a materialist approach to nature, one which is most convenient for this culturally created economic system, nature and the poor become unfortunate byproducts of consciously created structures, ones devoid of the responsibility that goes with such intended agency. From there, having duplicitously ascribed "nature" to these cultural creations, we ironically suggest that we must not interact directly in them in order to make them just. For, as the meta-narrative goes, *this* cultural creation (as with the previously mistaken view of nature) is mystically self-regulatory.

In short, when nature is seen as self-sufficient and self-regulating, and we are able to convince ourselves that certain cultural creations are in fact natural, poverty conflates as just another unfortunate byproduct of this "natural," unavoidable necessity within the deterministic logic of *laissez faire* capitalism. As such, this "unavoidable" poverty becomes institutionalized, rather than addressed, and thereby granted absolution. By first recognizing the false divide between culture and nature, there arises new onus on the observer to take responsibility as witness. No longer confined to passive observance, she can now engage a new *and* old aesthetic, making possible authentic responsibility to respond to the often reciprocal sufferings of nature and the poor.

PART II: SOMETHING OLD, SOMETHING NEW

Foltz details this "old" aesthetic in his work, *The Noetics of Nature: Environmental Philosophy and the Holy Beauty of the Visible.* In the introduction he articulates the ancient understanding of *Theoria*, one which is the root of the modern term "theory," the latter having primarily to do with specific operations of the discursive intellect. At one time, however, it meant much more than this. Foltz writes:

> Before the rise of 'theory' in the modern sense—and before Aristotle's celebration of what he called the *bios theoretikos,* rendered by Heidegger as 'the way of life of the beholder'—ancient *theoria* was much more than the metaphysical optics that Heidegger understood it to be. *Theoria* was contemplation, but not the observation that fixes in place, striving to 'entrap' and 'compartmentalize' its object, as Heidegger contends. Rather, ancient *theoria* was before this a 'mystical' seeing of the invisible within the visible . . .

Foltz continues:

> Yes, *theoria* is rooted in *theasthai,* the seeing that was once associated with the theatre. But ancient drama was far from merely 'looking' at what is 'present at hand' in Heidegger's sense. To be present at the theatre of Eleusis and watch

> the divine mysteries unfold, sitting at the very place where the invisible once emerged into visibility, and now seeing once again the invisible presented dramatically within the visible—this was not a process of objectification, but an event of participation, of taking part. *Theoria* was for the Greeks a special kind of 'experientially engaged' seeing, closely related to wondering. Nor was the beholder a curious onlooker.[18]

Whether or not this is a fair reading of Heidegger, the central point remains, that the perception of the invisible *by way of the visible* is the primary issue in what is termed here *noetics*. The Christian patristic understanding of the Greek word *nous* includes the rational intellect, but unlike the medieval scholastics, implies a more nuanced perception not limited by the confines of reason alone. Traditionally, the *nous* perceives effectively when seated in its spiritual center, the heart, where all faculties work in concert to "see." Here we have the patristic interpretation of Christ's words that "the pure of heart will see God," a heart which combines all faculties as one to grasp a depth of experience not available to the *nous* in isolation or when simply operating as the discursive intellect.[19]

What was previously invisible becomes visible through the change in the viewer and this renewed, *noetic* sight. The example of the ancient theatre as an unfolding process asks the viewer to become the participatory witness over detached observer, to hold the gaze until the invisible presence is revealed and perceived. This harkens to the logic of Yaweh leading the Israelites through the desert, asking them to follow an invisible God who will continually reveal the invisible presence in the visible, be it a desert cloud or a burning bush atop Mt. Horeb. Without focused, participatory openness toward the invisible—the clearing of a welcoming space anticipating its coming—and perhaps when tired by the difficulty of this struggle, God is reduced to a visible, finite object, one which requires no change in perception of the viewer to grasp; i.e. a golden calf, a form comprehensible, ordered and rational. The calf is complete unto itself, a "cultural creation of beauty," not calling the viewer to participation or openness to change, as it stands independent and whole, while paradoxically being an utterly dependent cultural creation that can be completely manipulated at will. But the calf is granted independent status despite being created by humans. And if it is beautiful it is a consoling and safe beauty, calling nothing from the viewer to grasp its literalism.

As with *laissez faire* capitalism, this cultural creation is given ontological independence, when in reality it remains utterly dependent upon its creators and practitioners. In both cases, the idols are deified by those who created them. The cultural creation remains a "thing" we can get our minds around and rest passively, over and against the dangerous, often chaotic beauty of the previous Theophanies of the unknown through nature, the latter of which are not our creation and require vulnerable, participatory witness, precisely through *not* knowing the other as object or as something comprehensible.

Again, we see this logic at play in Eastern Christian iconography, with its flat dimensions and specific sort of impressionism. The point of eternity (as one monk put it to me) is the viewer, the icon itself an immediate, *unrealistic* portrayal that calls forth the viewer to relational participation in a reality left intentionally messy, giving the rational intellect no place to rest in isolation. This is in opposition to the totalized, ordered form that satisfies the intellect, yet which does not ask the viewer to piece things together within herself to "see," making her a passive recipient of, in essence, a golden calf.[20]

An aesthetics of participation is engendered by the traditional form of the icon and better facilitates the *noetic* vision Foltz and numerous Church Fathers describe, a vision attached to the heart as the spiritual center of all the faculties working as one. Having reached this state of unity, the *nous* becomes the place of total vision, not reducible to the abstract intellect or didactic reason alone, the latter of which, when operating in isolation, is largely the problem.[21] For this old aesthetic to gain traction in a postmodern, technological age which often dogmatizes against any and all layered depths, a close look at what today passes for imaginative creation proves helpful. More specifically, the example below will convey an unconscious reduction of the imagination to dominant cultural forms, ones directly in contrast to nature and avoiding such layered depths in the process.

PART III: VIRTUALLY REAL

In a work by British philosopher Glyn Daly entitled, *Conversations with Zizek,* contemporary Slovenian thinker Slavoj Zizek discusses a well known psychic malady, that of the compulsion to cut oneself. As Zizek notes, the most popular exploration of this phenomena suggests that it arises out of the need to "escape" the real, to feel something other than that person's presumably miserable current situation. Under this rubric, cutting provides an altered consciousness and can become an addiction of sorts by providing this momentary "escape" from drudgery. This might be described as an attempt to escape the pressure of being oneself, the cutting a release of tension to temporarily free oneself *from* oneself.

However, Zizek offers another interpretation (one clarified further by contemporary thinker Frank Seeburger), that the person in question does not feel reality enough; the compulsion in this case a direct response to the person *not* feeling herself real, the physical pain and drawing of blood breaking *into* reality rather than away from it.

To bolster his point, Zizek refers to a popular David Fincher movie, *Fight Club.* After departing from most critics who either saw the movie as a glorification of violence, *or as a critique of the glorification of violence,* Zizek writes:

> Few saw in the film something which I think we should have the courage to accept. That is, the emancipatory dimension of this self-beating, and that, in a way, we need to take a risk through this kind of violence. When we live in a virtual isolated space, every reconnection with the real is something shattered; it is violent. This is why, today, cyberspace virtualization is necessarily supplemented by different forms of the 'return of the Real'—from politically 'regressive' activities like new racisms to bodily mutilations and so on—there two sets of phenomena are strictly correlative.[22]

The crux of the issue is this "cyberspace virtualization" and its need for supplementation. Unlike Eastern Christian icons and Van Gogh's impressionism, virtual realities remain too close to the empirically real to call the viewer into active participation by way of changed perception. By themselves, virtual realities do not facilitate a return to the real, but lead instead to the "regressive activities" Zizek points out as suddenly attractive (or even necessary) in some contexts to facilitate reentry to the fully real.

Along these lines, and in reference to *Fight Club,* Seeburger observes that main character Ed Norton, as well as his imaginary opponent, Brad Pitt, are not an escape from reality, but once again, a reentry, as the phantasm of Pitt is an imaginary projection of a different side of Norton's own self. Indeed this is Norton's way back to integration of the real, as with one interpretation of the one who cuts herself. In reference to Zizek's interpretation of the film, Seeburger writes, "Thus, Zizek helps us reverse the formula in accordance with which fantasy is an attempt to escape from 'reality.' He helps us to see how, far more significantly, 'reality' is an attempt to escape from fantasy."[23]

In this case the "flight from fantasy" leads to a desperate return to the *reductively* real or the immediate-empirical (as if *that is* the real), and is at the same time a flight from the imagination and the *telos* of the real as that which can include the invisible presence beyond the immediate-empirical. The variety of virtual worlds, the varied imitations of immediate empirical surface perceptions, are in fact a bypassing of the imagination while promoting themselves as its creative products. Virtual realities, whether through video games or the ever progressive special effects in movies, are still too similar to the empirically real, such that the imagination is not needed, the "piecing together" of what is viewed and, more importantly, the piecing together of the viewer herself. By mistakenly seeing virtual realities as deep imaginative projects, we quickly assume that such projects are "all there is" or, as with the golden calf, all that the imagination can muster, nothing more than a literal mimesis of the empirically real.

Unlike the theatre of Eleusis or the Hebrews wandering the desert, we no longer *actively* wait for the invisible; the virtually real presents us a slightly altered reality, one comprehensible and whole, and one in which to passively rest through no effort. Thus the attempt to break out of the immediate-empirical by creating virtual worlds only bypasses the imagination while thinking it in use. This in turn leads back to "the real" as that which is most immediate and does not call the viewer to participate in its fullness or that which includes a presence beyond the visible.

The imagination's dormancy as a result of its being bypassed by an entry into the "faux fantasy" of virtual worlds, during which the fullness of reality as including the invisible continues to lurk beneath the empirically real, may require the violent breakthroughs Zizek mentions in order to enact a completeness of perception previously bifurcated. Seeburger writes, "The liberators in our dreams appear to us as monsters, so we flee from them as we can into 'reality.'"[24] Here too 'reality' remains in quotes to mean that this fleeing is into a new reductive reality of the immediate-empirical, suggesting that turning *toward* the monsters (as Ed Norton turns toward himself and becomes aware he is beating himself at the end of the film) would facilitate return to the fully real.

Again, this escape into a reduced reality or the "flight from fantasy," suggests the need for a violent breakthrough to fullness, that one needs to turn *toward* the monsters, a turn which would expand the previous reduction of perception in the newly integrated psyche. Van Gogh's *Starry Night* can appear at first as an invasion of inner and outer space that almost suffocates the viewer in a vivid (psychotic) obituary, before discharging one into the mystery that night is typically lost these days in mundane urban blight. For Maximus and the Transfiguration, the expansion of perception in the viewer is where the change occurs, now able to see what was always there, but which required the fullness of being, the gathering of the faculties, to participate in the depths of the real. Without the cultivation of such vision, the attempt at entry into the real may need to be a violent one to overcome and shatter the accepted split between culture and nature.

Part of the logic, then, of Van Gogh's desire for the "ugly" can be seen in that the "ugly" provides the necessary violent offense to cultural beauty to awaken the viewer into the fully real and change her perception. Such offense calls her to participate, to wait for the invisible to piece together the real, the process of piecing together part and parcel of reality itself. In this sense, the culturally ugly can bring us into the fullness of natural beauty, in distinction from a cultural beauty which finds itself complete. This is over and against nature as well, as this completion seeks a conscious distancing from nature, a "fleeing from fantasy" into the reduced real, the idealized and ordered form; i.e., the golden calf.

PART IV: CULTURAL ENVIRONMENTALISM

The idealized and ordered form is today so pervasive that it infiltrates conceptions of beauty in the most ardent lovers of nature. The typical response to the current environmental crisis rightly points out a widespread disregard of beauty found in nature. Yet in doing so the inherited divide between nature and culture remains. In other words, the typical response points to nature's beauty by displaying pristine meadows and snow-capped mountains, a beauty which fits nicely within culturally ordered norms, ones pleasing to a created cultural aesthetic, even though such mountains were once seen as "repulsive" and "bleak" by other cultural norms. If this is the extent of natural beauty, there is no need to look into or beyond the immediate to access the fullness of the real. A sleepy, cultural aesthetic grants surface validation. This kind of beauty, certainly capable of bringing good feeling, remains consoling at bottom. It is a culturally accepted, ordered perfection that asks only that we put ourselves passively within its terrain to enjoy and receive consolation.[25] As such, this approach reinforces the cultural construct of what is to be considered beautiful, Ansel Adam's photos fitting nicely within what is accepted by the *nomos* to be "natural beauty." As a result, cultural norms of beauty reign even when attempting to retrieve natural beauty.

Most problematic, as Zizek points out in the documentary film *The Examined Life*, is that this approach mystifies the environmental problem by presenting nature as an idyllic, self-regulated mechanism in separation from us, even now amidst obvious man-made damage. It lulls the viewer into believing that nature cannot be destroyed just as a projected cultural beauty (onto nature) soothes us to sleep. Be it a tightly manicured golf course or a well-groomed ski slope, nature becomes an object for use and not an "other" in which we have a dynamic, contemplative relation in order to perceive the real, much less a divine dwelling for the awe and risk of theophany.

In this case beauty loses all paradox or complexity which might trigger the imagination of the viewer through the sort of violence that is in fact part of nature's deeper, natural beauty, one which demands awareness in the perception of the viewer, to see more than the aesthetic surface, a violence revealed to anyone willing to wait, not unlike the waiting by the Israelites or the waiting at the theatre of Eleusis. In this sense, "natural beauty" is a paradox of immanence and transcendence, *nomos* and *physis,* left unresolved by rational dialectics and known primarily through active engagement and even detached expectation. And yet the *lack* of paradox in the immediacy of cultural beauty even impacts which endangered species gain attention, the obvious, powerful surface beauty of the tiger verses the culturally grotesque beauty of frogs, as just one example of which animals are able to gain financial support against extinction.

Again, Zizek responds to this idealization of nature by emphasizing that nature is a history and series of catastrophes, and that by only showing pristine meadows in the hope of moving us to action, the opposite actually occurs. Promoting such culturally accepted beauty reinforces the culture/nature split by solidifying the deep-seated fantasy that this seemingly independent, aesthetically pleasing world will never be lost.[26] Once more we have the difference between the reductive real as a flight into the immediate-empirical, as opposed to the fully real, or that which calls out the viewer to change and perceive the invisible presence through that change. When the portrayal of nature points to itself as complete, as if there is nothing more to it than what is passively and immediately seen materially, we turn nature into another rational idol, the intention of the perceiver having everything to do with what is perceived; i.e., what you see is what you get. Materialist expectations engender materialist findings.

Speaking from a garbage dump and surrounded by thrown out computers and refrigerators, it would seem that Zizek wishes to break us into the real when he makes the surprising claim that, rather than showing idyllic scenes from nature, we must instead become ever "more artificial," perhaps employing the rationale of fight club, that only such a violent turn will awaken us to what is real and the realization of the mutual interdependence between culture and nature.

While Zizek is quite right in his critique of certain idealizations of nature as counterproductive, it seems unnecessary to become more artificial to spark the violent uprising that will break into the real (if that is his motive). The critical consideration is that nature itself provides ample opportunity for this awakening in the viewer, especially in light of Zizek's own characterization of nature as a "history of catastrophes." When holding the gaze, the violence and danger within nature itself breaks us through surface layers by not letting us rest within a cultural aesthetic or false consolation. But again, this requires an active participation, a waiting to "see." The participant must beware, engage the unknown and, to a certain degree, engage being out of control.

PART V: EROS AND IMAGINATION

Having lived well before the virtual age, Van Gogh articulated a certain disdain for a method that directly applies to our time. Biographer Ingo Walther writes of Van Gogh, "The dogma of antique beauty repels him. This dogma reduces the individuality of art to a mere mechanical copying."[27] In distinction from such copying, we see in Van Gogh's early work the strategy most indicative of his unified vision by way of an unusual *lack* of stress on any one object within a painting. Through a certain subtle blending, even at times with tones *more* subtle than those found in nature, Van Gogh found this approach to be a point of departure from other artists. Walther continues:

> The question arose how one could convey pictorially the changes in color of an object which becomes apparent when one distances oneself from the object or observes it in a different light. The solution finally agreed upon was to tone down the actual color of the object, its natural color, in order to fit in with that of the dominant monocolored tone of the painting as a whole. No longer was the singularity of the object stressed, but its appearance within the context to which the picture was dedicated. This tendency towards a separation of color within the painting from the concrete object in reality was the most important step towards a definite autonomy. [28]

Along these lines, one could argue this was the case throughout his whole career. To consider one example, that of "The Pink Peach Tree," a painting which displays his change in pallet to an explosion of color, he nevertheless brings the sky down into the tree, into the bark, even though that was not how the eye objectively saw it. Thus while drab tones were one strategy to call for creative unity in the viewer, the goal remained this unified vision toward the experience of nature, even when overwhelming the viewer with bright colors. By moving away from what the eye objectively and literally sees, once more the viewer is asked to piece together the whole, and in doing, enter the scene loosely depicted. Most of all, by *not* conveying nature as a unified whole unto itself, the viewer either imaginatively situates herself within the painting or turns away. In either case, she is forced to act, unlike the passive response of a virtual reality that rotates upon itself and the empirical surface. Most of all, in order to keep the gaze, one recollects one's faculties; in Christian patristic parlance, the *nous* descends into the heart in order to "see."

Keeping this in mind, Van Gogh's criticism of the nude *qua* nude seems more pertinent today, given techniques of airbrushing culturally perceived flaws into another virtual reality. Here the imagination is once again bypassed, reducing the reality of the erotic from a relational phenomenon between particular people, to a Pavlonian response between an idealized *techne* and a passive recipient. In this case the nude is a virtual reality reduced from *physis* to *nomos,* ever frustrating the viewer by the sudden inability to reach the real while feeding the idolatrous fetish for the cultural creation, one that cannot be satisfied precisely because of the "flight from fantasy" which has sedated the active imagination of the viewer. Again, in this case the viewer remains passive, the airbrushing technique solidifying the real as the empirically immediate and yet, through the airbrushing, it is not even that. Here too, only a sort of violence potentially breaks through for the viewer. Such violence is all too often against the very persons who fail to "measure up" to the cultural fetish.

The ability to see *noetically,* what Foltz calls the ability to "see the invisible in the visible," fits well with the description by Maximus of the Transfiguration, an event made possible by the change of perception in the viewers after responding to the call to participate in "the Way." A similar phenomenon is expressed throughout Van Gogh's work, the attempt to convey nature in its truth, as relational icon which continually calls us forth: that to see into the painting requires a change in the viewer from passive to active, the painting itself a "window" through which the real passes, meaning that the viewer either merely looks at the window, or sees through from the window, participates in the invisible and moves forward.

This unified, participatory vision has never been more important than it is now, whether it is a matter of overcoming the denial of the cultural impact upon nature, or how this logic of separation and self-regulation extends to culturally constructed institutions. The poor who Van Gogh depicted, who historically have lived in the worst environmental situations, are often seen as unfortunate, "natural" byproducts of an independent, self-regulating Capitalism, an ironic mystification that ignores the latter as a cultural creation and by doing so, avoids responsibility for its care or improvement. Even more mystifying, this avoidance is seen as "natural." As such, the cultural responsibility for environmental degradation, as well as the poor who feel the deepest and most direct impact of such degradation, are granted absolution. To tamper with either would be a cultural sin by tampering with what is now seen to be natural, the word "natural" come to mean an unavoidable destiny which is simply out of our hands.

The contemporary viewer's expectation of nature, formed in part by the inherited, presumed divide between culture and nature, has everything to do with what is perceived there, a certain materialist expectation laying to rest the role of the imagination, a role which historically (whether in Maximus's Eastern monastic context or Van Gogh's artwork) calls the viewer to piece together the invisible presence at work within visible nature while simultaneously piecing together himself. By extension, similar presuppositions have everything to do with tragically low expectations of culturally created institutions, especially with regard to lessening if not eradicating poverty through them.

Once systematized, a certain corruption of beauty as an idea creates its own capacity to betray itself, arguably a more insidious phenomenon than a culturally created morality in its ability to quietly justify a hierarchy of life, one that goes against our true nature as active, relational beings who see into the real and beyond the empirically-immediate. When the attempt is made to create beauty over and against nature, or when we see *techne,* nature and culture as independent of each other, we are in essence "kicking against the pricks." To exercise the capacity to enter and act within a beauty that is dynamic and transformative is to be aware as viewer that one is never completely seeing the full reality of the present. Part of the unceasing gift of the present is that it is always partially denied, partially hidden, and asks us to wait and act to piece together both ourselves and the real.

BIBLIOGRAPHY

Arendt, Hannah. *The Human* Condition. Chicago: University of Chicago Press, 1958.

———. *The Life of the Mind.* New York: Houghton Mifflin Harcourt, Inc., 1978.

Berleant, Arnold. *The Aesthetics of the Environment.* Philadelphia: Temple University Press,1992.

Daly, Glenn. *Conversations With Zizek.* Cambridge: Polity Press, 2004.

Drenthen, Mark and Keulartz, Joseph. *Crossing Divides and Breaking Ground.* New York:Fordham University Press, 2014.

Foltz, Bruce. *The Noetics of Nature: Environmental Philosophy and the Holy Beauty of theVisible.* New York: Fordham University Press, 2014.

Louth, Andrew. *Discerning the Mystery.* Oxford: Clarendon Press, 1983.

Nietzsche, Friedrich. *Thus Spoke Zarathustra.* New York: Oxford University Press, 2015.

Palamas, Gregory. *The Triads.* New York: Paulist Press, 1983.

Parson, Glenn. *Aesthetics and Nature.* New York: Continuum Publishing Group, 2008.

Seeburger, Frank. *The Open Wound: Trauma, Community and Identity.* CreateSpace Independent Publishing Platform, 2012.

Stead, Dom Julian, trans. *The Church, the Liturgy, and the Soul of Man: The Mystagogia of St. Maximus the Confessor.* Still River: St. Bede's Publications, 1982.

Taylor, Astra. *The Examined Life.* New York: Zeitgeist Films, 2009.

Tralbaut, Mark Edo. *Vincent Van Gogh.* Victoria: Abebooks, 1969.

Walther, Ingo. *Van Gogh.* New York: Taschen Press, 1999.

NOTES

1. Law here means prescriptive, human law, in distinction from "natural law," or that which is prior to empirical human endeavor, and in a sense stable before and apart from such endeavor.

2. Even with the pre-platonic philosophers like Heraclitus, who clearly saw a fluidity in the natural world, it was a fluidity that remained mechanistically independent of human interaction and, in that sense, set in place.

3. This notion of *noetics* comes from the Eastern Christian patristic use of the term *nous,* loosely translated as mind, but which connotes a certain cohesion of vision as the mind descends into the heart and the person is able to see again as a unified person. This return to full sight is in response to the "scattering" impact of the Fall, the latter of which has led to a compromised vision and perception of God and, in turn, of God through nature.

4. Maximus, *Ambiguum,* 10 (PG 91 1127 D-1128C), trans. Andrew Louth, 108-109.

5. *Ibid.* It is worth mention that the "deceiving senses alone" is a reference to the physical senses trying to perceive truth "alone," which they could only do with material presuppositions. For Maximus, whether it was the senses alone or the intellect alone, the point is that to perceive "noetically" is to have all faculties working as one and in concert with each other and never alone. According to Gregory Palamas, the faculties "working alone" is the direct result of the Fall, and that to bring them back together is one of the goals of prayer and the ascetic life.

6. It is important to grasp that when Maximus talks about "the deceiving senses alone," the "alone" portion needs emphasis. This is not merely a warmed over Neo-Platonism as much as it is an articulation of the need for all faculties to be working in concert, senses with the intellect and imagination. The "senses alone" is the materialist presupposition, that all that is available is sensorily perceived; hence the deception.

7. German Political Philosopher Hannah Arendt alludes to Kant's awareness of this double life of the eye when she comments, "It is, therefore, indeed always the same subject that is both a member of the visible and the invisible world, but not the same person, since...what I as mind think is not remembered by me as man, and, conversely, my actual state as man does not enter my notion of myself as mind." And Kant expressed in a strange footnote of a "certain double personality which belongs to the soul even in this life. (Arendt, *The Life of the Mind,* 45).

8. Foltz, Bruce, *The Noetics of Nature,* 104.

9. While the indigenous aesthetics of nature was certainly (in part) a creative product, it was one without the inherited culture/nature divide that eventually led to a much needed "return to nature" in the modern West.

10. Though beyond the scope of this paper, the issue of negative capability and how much a poet can place consciousness into nature without being accused of anthropomorphism is debatable and perhaps a reason why prose nature writing has been more widely accepted to instruct the public in general.

11. Foltz, Bruce, *The Noetics of Nature,* 42.

12. This notion of humans as "co-creators" with God is traditional view of the Christian East.

13. Arendt, Hannah, *The Life of the Mind,* p. 178.

14. Arendt, p. 178.

15. This is akin to Heidegger's insistence on "doing philosophy" and not "learning about" philosophy.

16. Foltz, *Noetics* p. 117. While this is its own project, the problem with ontotheology for our purposes is that it precludes allowing the icon to work as a facilitator to anything more than the highest being among beings. In this sense, icons are not really icons.

17. This will become more important as we consider Zizek's critique of the environmental movement and its tendency to idealize nature in such a way that perpetuates the inherited culture/nature/ divide.

18. Foltz, Bruce. *The Noetics of Nature,* p. 3

19. It should be clear that this is no moralistic purity, as those who recognized Jesus as the Christ included a prostitute and a tax collector, one possessed by demons and another with "five husbands."

20. More realistic portrayals tend to convey something literal, that the depiction is of something that took place *in the past*, less able to facilitate the ongoing presence of that event now. The Eastern icon, by employing an unrealistic, impressionist style, avoids "the fallacy of imitative form" and better facilitates the present experience of presence by *not* attempting a copy of the past.

21. Gregory Palamas suggests that the Fall "scatters" our faculties. As a result we fail to "see" God. Prayer, fasting, almsgiving, etc., unifies us spiritually so that we can better "see" again.

22. Daly, Glen *Conversations with Zizek.* pp. 118-119.

23. Seeburger, Frank. *The Open Wound,* p. 20.

24. Seeburger p. 20.

25. This is an implicit critique of Kant's understanding of beauty as consoling.

26. Zizek in documentary film, *The Examined Life,* by Astra Taylor. Zeitgeist Films, 2009.

27. Walther, Ingo, *Van Gogh* p. 17

28. *Ibid.,* p. 19

Chapter Three

The Standpoint of Transformativity

Re-envisioning Science, Nature, and the Self

Mark W. Flory, Ph.D.

> For this reason I bow my knees to the Father of our Lord Jesus Christ, from whom the whole family in heaven and earth is named, that He would grant you, according to the riches of His glory, to be strengthened with might through His Spirit in the inner man, that Christ may dwell in your hearts through faith; that you, being rooted and grounded in love, may be able to comprehend with all the saints what is the width and length and depth and height— to know the love of Christ which passes knowledge; that you may be filled with all the fullness of God. (Ephesians 3: 14–19)

I. THE ART OF REINTEGRATION[1]

In Paul's Letter to the Ephesians, he provides us a sort of précis of this essay. In the first place, he provides a model of the first stage of the system of spiritual practices, namely purification: the bowing of the knees, and the humble invocation of Jesus Christ as creator and redeemer. He appeals to God to transform the inner being of his audience members. Indeed, the phrase "that Christ may dwell in your hearts through faith" suggests strongly the practice of "bringing the mind into the heart" associated with Eastern Orthodox Christian Hesychastic spirituality (about which more, below). These purificatory conditions, Paul suggests, are the conditions for comprehension of "the width and length and depth and height"—in other words, the nature (or *logoi,* essences) of created being. Paul calls upon us to undergo nothing less than a total transformation of our nature, and it is this transformation of the conditions of our natural being that will make possible, paradoxically perhaps, the fulfilment of our knowledge. It is this mysterious progression that this paper seeks to illuminate.

For many today, science provides the most important explanatory system, not just at the natural level, but, usually implicitly, at the metaphysical level as well. Within the rather mind-numbing mud-slinging of the so-called science and religion "debate," defenders of religion often point to these metaphysical assumptions. Their aim is not so much to correct science, as to condemn it, thereby causing the inevitable, positivist and reductionist backlash on the part of scientists. In this paper, I want to attempt to provide a broader context for the debate, and for science. Science, after all, is only one way in which we come to know nature (our own nature, the nature of other beings, and the nature of nature itself).[2] Knowledge of nature also plays a vital role in other systems, such as systems of spiritual practice, environmental philosophies and theologies, aesthetic theories, and art itself. What might these other ways of looking at nature provide in terms of a re-contextualization and re-envisioning of our relationship with nature?

I will argue for a dynamic understanding of science and religion that allows for the possibility of *reciprocal transformations* between and within each of these pursuits. In the first place, I will present a theory of practice based on my own study and practice of Eastern Orthodox Christian Hesychastic practices. This theory, unlike most theories regarding science or spirituality, begins not with metaphysical claims, but with the transformative mechanics of spiritual practice itself. Metaphysics are not ignored—in fact, according to my theory, a cosmological vision is essential to the functioning of a transformative system—but I am attempting to avoid the pitfall of many theories that begin with reified metaphysical concepts (such as self/other, human/divine, subject/object) or very questionable historically-based generalities.

To provide context for the theory, I present examples from Plato, Kierkegaard, and the Hesychastic system of Gregory Palamas. I utilize these examples to re-contextualize science and the study of nature according to the exigencies of transformativity: the mutual and perpetual transformation of self, others, system, and nature itself.

A. The Theory of Practice

A practice comprises, in the first place, a reciprocally transformative cosmology and anthropology, along with the methods and exercises that produce the kind of person which the cosmology valorizes. The symbols in which these reciprocally transforming elements are expressed are over-determined, both in virtue of their semantic multireferentiality, but also because of their transformativity. That is, they each and all transform each other, and are transformed. The symbols are dynamic, organic reifications of the transcendent imagination.[3] This imagination is the conditionless condition for symbolism *per se*. In the transcendental imagination, *mythos* and *logos* are neither united nor divided, but they reciprocally transform one another.[4]

The theory of practice is itself multivalent and multidisciplinary. Variations on the theory are found in several of the social sciences.[5] I have developed my own version of practice theory based on Eastern Orthodox Christian Hesychastic spiritual practice.[6] I bring that theory into transformative conversation with other systems of spiritual practice and philosophical transformation. I have found that as I analyzed the mechanism of Hesychastic practice, the theory of practice I developed related complexly to other theories of practice, which were based on different systems of practice. The basic premise underlying all these approaches to practice, however, is that such phenomena as knowledge, meaning, human activity, science, power, language, social institutions, and historical transformation occur within and are aspects or components of the *field of practices*. The field of practices is the total nexus of interconnected human practices. The 'practice approach' can thus be demarcated as all analyses that (1) develop an account of practices, either the field of practices or some subdomain thereof (e.g., science), or (2) treat the field of practices as the place to study the nature and transformation of their subject matter.[7]

In many presentations of practice theory, however, the analysis occurs at the level of relationships or interconnections within a "field." I will argue throughout this essay that this is the wrong level of analysis of such systems, and that at such a level we constantly risk mistaking our examples for our categories (reductionism), and *vice versa* (reification). Reductionist theories of science and other systems typically take specific historical occurrences or conceptualizations as if they demonstrated something about the essence of the system. For example, to say that the Aztecs practiced cannibalism because they lack protein in their diet is a materialist reduction of the complexity of Aztec motivations to a fact (which turned out not to be a fact at all) that was, at best, merely one motivating factor. As we will see, below, some theories of science commit a similar reduction: they read the essence of science on the basis of isolated historical facts or concepts.

On the other hand, in our attempt to attend to the complexity of systems, and even moreso to their transformativity, it is very possible to resort to a reification of basic concepts. In such a case, we risk taking the concept for the thing itself. As we will see, complexity theory reminds us that "science" and "religion" are not only complex, but they are reifications of a multitude of factors (and very fuzzy reifications, at that). I will use Evandro Agazzi's "system-theoretic" approach to systems to attempt to move beyond the shortcomings of the complexity thesis, and I will argue that while Agazzi's approach generally operates on a properly general level of analysis, he, too, places conceptual limitations on transformativity.

Conceptualizing a practice as a reciprocal transformation between cosmology, anthropology, and *praxis* provides the properly general level of analysis—a claim I will demonstrate by reference to Plato, Kierkegaard, Hesychasm, philosophy of science, and other means.

As soon as we assume this more general (*meta-*) level of analysis, we see that many of the regnant pieties of historians of science must be challenged. For example, a common reading of the history of science tells us that the emergence of science coincides with the decline not only of any given metaphysical system or systems, but of metaphysical explanation itself. However, a metaphysic never exists independently of an anthropology and the exercises that realize the system's *telos*. When the mutual link between the metaphysic and the activities changes, becomes tenuous, and breaks down, transformativity is compromised.

On the other hand, the mere correspondence of a metaphysic and activities does not necessarily imply their transformativity. Often, historically, a metaphysic seemed to support, and to be supported by, activities (individual, communal, social, even political and ideological activities). But when the mutual support of metaphysic and activities becomes tenable only through the exercise of institutional power, ideological axe-grinding, and fear of change, transformativity is at a premium.[8]

We will see historical examples, below, of metaphysical systems that have outlived their (mutually confirming) practices, and systems of practice that have become unmoored from their metaphysics.

To determine the transformativity of a system we need to examine the dynamics of that system: the multivalence of its symbols, the reciprocally transforming relationship of practitioners and practices, and its open-ended teleology. When we examine these dynamics, we discover a strange paradox, one that makes all the difference.

In certain practices, the activities of subjectivation[9] can produce transformations that go beyond natural conditions.[10] These conditions include the intentional (subject-object) structure of consciousness; the orientation of consciousness in space/time; the experience of unitive knowledge; and supersensory powers.[11] The transcendence of natural conditions is the beginning of the paradoxical processes of enlightenment (or illumination), an experiential awareness of the nature of reality, and the nature of one's own mind.

In addition to the paradoxical nature of the transformative process, many of those who have experienced this transcendence of natural conditions make paradoxical claims for this experience. First, it is said to be ineffable—but next, it's said to be productive of knowledge superior to natural knowledge. If it is ineffable, how can it be superior? If it is superior, how can it be shown to be so? First, it is said to exceed to limits of human conceptuality, next, it's said to be necessary in order to comprehend existence itself. In the next section, we will consider two famous examples of this paradox: Plato's cave allegory, and Kierkegaard's Knight of Faith and teleological suspension of the ethical. In both cases, I will argue that the seeming diremption at the heart of these progressive systems tends to disguise a subsequent (and consequent) movement in which *the transformed becomes the transformer.*[12]

Traditionally, the venue for such transformations has been religion, though in modern times the religious systems of transformation (spiritual practices, spiritualities, spiritual exercises) themselves have been lost or at least historically disrupted.[13] However, today science has increasingly made its claim to providing the dominant paradigm/grand narrative for our time. Often, science (to speak abstractly now of a concept which we will have to problematize later) has claimed that its success (at prediction, discovery, and invention) justifies the status of its findings as knowledge, and/or that this success has occurred independent of any *a priori* cosmology or metaphysic. The result, Bruce Foltz writes, is that we live in

> a world that has elevated the *methodological* materialism of modern natural science and technology into an all-encompassing *metaphysical* materialism, a hegemonic materialist *Weltanschauung* that effectively rules public discourse and decision-making. For modern materialism is the metaphysical correlate of epistemological scient- ism, i.e., of the view that sees scientific knowledge as not just the highest, but the *only* legitimate knowledge of nature. Methodology prescribes reality, delineates its profile and parameters beforehand. Or as Heidegger put it, science is for modernity 'the theory of the real.'[14]

Even where scientists and defenders of science have recognized the absurdity of claiming that science is preconditionless, however, they rarely examine the implications of the conditioned, relative nature of science.[15] The theory of practice just outlined provides a tool for such an examination (see below, section IIA).

The theory of practice provides a comprehensive symbolic system (a cosmology) that we can engage through exercises and activities that produce in us the kind of transformations that the cosmology upholds. A system of practice, then, is not a mathematical or scientific or logical system, nor is it necessarily linear or unidirectional. Engagement of a system of practice is, ideally, a total participation. However, the system itself cannot attain its own ends—a transcendence of the conditions of the system itself is necessary to fulfill those ends. Transcendence, therefore, is the precondition (a necessary but not sufficient one) for transformativity to occur.

In the next section, I will provide more details about the theory of practice, and provide two examples—Plato's cave allegory, and Kierkegaard's teleological suspension of the ethical—of the movement of transcendence, and of the transformative reintegration that this transcendence makes possible.

B. Transcendence and Transformativity

The system of practice, as I have outlined it, provides a comprehensive context for the transformation of nature. Such transformation effects one's own nature, the nature of other persons through our agency, and nature in the sense of natural being itself. However, these effects are the result of a paradoxical process that includes the transcendence of nature and natural being. This transcendence implies that our conceptuality and language will also be transcended. Why is such transcendence necessary for full transformativity to take place? How can we become agents of transformation if we have transcended the natural conditions of conceptuality and language? In order to answer these questions, I present the following examples.

Plato has been labeled a hater of art (or at least poetry), and while the accusation needs nuance, there is one aspect of his aesthetics that might earn him that label. If we consider the implications of the Cave Allegory, the process of exiting the cave requires an extensive *paideia* that in the first place, through gymnastics and music, leads to self-control over one's appetites, emotions, passions, and lower cognition ("monkey mind"); and later, via mathematics, logic, and scientific reasoning leads to transcendence of one's bodily nature altogether. This last stage also includes the practice of Socratic dialectic, which eliminates the even more uncontrollable element of our ego-bound belief in our own opinions.

Plato is clearly stating that true knowledge, knowledge of the Forms, requires the total transformation of human nature, to bring it into accord with the Forms. The objects of the senses, and all the "knowledge" based thereon, are not illusory, but they are deficient. Here, Plato gives us the metaphysical justification for his concerns about art: art titillates the senses, it excites the emotions, it fosters distraction and the enflaming of false and futile desires (which can result in social unrest). Art does not merely represent nature—it is an imitation of an imitation. Nature itself is an imitation of the Form of the Good, because it is itself a mixture of illusion and truth.

However, it is also possible to see in the Cave Allegory the possibility of a bridge between the "two worlds." As the prisoner exits the cave, she is at first blinded by the brightness of the Sun. In order to allow her eyes to adjust to the light, she first views the reflection of the Sun in pools of water, and the leaves of trees. She then raises her eyes up to those objects that reflect the light of the Sun—astronomical objects, the moon and stars (but perhaps metaphorically other things that, moreso than most natural objects, reflect the truth). This progression suggests a continuation of the *paideia* begun inside the cave, and gives us, perhaps, the seeds of a Platonic "natural theology."

The pools of water, the leaves of trees, and the astronomical objects *reflect* or imitate the Form of the Good. They are superior to other natural objects because they reflect the Sun more clearly, but they are not the Sun itself.

Plato also suggests a further stage of this enlightenment process, when the escaped prisoner (the philosopher) returns to the cave to try to awaken the prisoners from their ignorance, egotism, and delusion. This could also perhaps suggest that the philosopher, after transcending the limits of natural cognition, after having transformed their body, soul, and spirit to become like the Form of the Good, has to reclaim these "lower" aspects of her nature, has to reintegrated those faculties that were previously transcended.

If this final stage of the process is ignored, we have the strange spectacle of an enlightenment that takes one beyond nature, without the means to effect any change in that nature—not transformation, but transcendence *per se*. Not only this, but we would have the situation described in Plato's *Ion*: Socrates forces Ion the rhapsode to accept that his art is one of divine possession, but that therefore Ion cannot claim any knowledge of the specific arts that Plato mentions. (*Ion*, 542A) Socrates points out that Ion's ability to enact Homer's words such that he and his audience are transported does not provide knowledge.

In regard to this problem, the Cave Allegory provides a possible solution. While the cumulative effect of Platonic *paideia* is to transcend natural conditions in enlightenment, the Allegory suggests that this would be an incomplete process without the return to the cave, the reassimilation or reintegration of those natural conditions. Contrary to the common (uncritical) claim, Plato was not advocating dualism, but a transformation process that, by means of gradual transformations, frees the subject from natural conditions, which in turn enables the subject to become an agent of transformation in nature and society.

We find a similar common misconception in interpretations of the work of Kierkegaard. Kierkegaard claims that his three stages of subject-formation—the aesthetic, the ethical, and the spiritual—are not strictly progressive. That is, he tells us that they are cumulative, rather than serial, in nature. However, the spiritual stage so subverts the conditions of the other two stages that it isn't even in the same sequence as the other two. The teleological suspension of the ethical connotes the transcendence of natural ethical categories, and the absurdity and impossibility of God indicates the transcendence of natural knowledge.

If Kierkegaard stopped there, then Derrida's warning in regard to the impossible would be apt: the absolute other could be a monster. Derrida presents this danger in analyses of a range of modes of impossibility (that is, if impossibility could have modes!): the pure gift, messianism, hospitality, and others. In each case, the purity of the concept is directly proportional to its unknowability and impossibility. In response to Jean-Luc Marion's attempt to domesticate these modes of impossibility, Derrida claims to be protecting the otherness of the other.

But if these modes of impossibility are truly impossible, there would be no means of instantiating them, describing them, presenting them. Terry Eagleton famously criticized Derrida's philosophy precisely for its impossibility: in the absence of any actual social or natural mediation, it not only could not be rendered in social praxis, it could not aid in the resistance to forces of social repression and alienation.[16]

Does Kierkegaard's spiritual stage also become so impossible as to be useless, or worse, complicit with the *status quo*? Is Kierkegaard seriously suggesting that the transcendence of ethical law is an end to be sought? In the first place, of course, Kierkegaard does not make the mistake of asserting that the spiritual stage is a universal goal (i.e., a goal for everyone). Only some few individuals, perhaps, will ever achieve it. In the second place, no one will ever achieve it on her own power—it is not a goal that can be achieved by striving for it. Rather, it is God who chooses the individual, not the other way around. The "leap of faith" is not so much a leap, as a being dragged (and we must remember that the prisoner who escapes the cave does not choose to do so, but must be forcibly dragged up the long ascent).[17]

But this does not lessen the impossibility, or make it more instantiable or less dangerous. The "knight of faith" is akin to Socrates' interpretation of Ion. Whereas the knight of infinite resignation—the ethical hero—veritably shines with goodness, Kierkegaard describes the knight of faith as an everyman, indistinct in every way from his fellow citizens, except that he responds to God's call, even when that call beckons past the point of absurdity. Commentators, in attempting to explain these strange claims, generally either reject Kierkegaard's scheme as incoherent, or find some ways of mediating them. Candidates for mediating God's relationship to human beings included an "ethic of care," forgiveness, interdependence, friendship, trust, faith, and love.[18] Notably, all these mediators relate to relationship. However, relationship is thought in terms of conceptual relativity or dialectic. If we take Kierkegaard's impossibility seriously, these attempts to mediate the absolute difference between humans and God undermine the very conditions they purport to explain.

However, an aesthetic of nature, informed by the theory of practice offered above, suggests a continuity of nature and super-nature, of the human and divine. This continuity does not deny that a qualitative difference exists between human and divine, but that that very difference is the condition of the possibility of transformation. Transformation, in turn, as it becomes more and more continual and reciprocal (transforming subject *and* object, human and divine) does not transcend nature, but transforms nature *from within*.

Therefore, while in the processes of enlightenment, the conditions of natural subjectivity (and cognition, and *aesthesis*) are transcended, they are not merely abandoned. Rather, through *theosis*—the Hesychastic term for the practices of perfection[19] —they are transformed, taking on new context and meanings.

This transformative process describes an ascent that, like a mountain peak, and has various approaches. Most obviously, I am referring to art and aesthetics. In this case, the conditions of natural sensation and perception are transformed, revealing divinity within nature. But the same process applies also to science, in the sense given to that word by Hellenistic and Christian authors. In this sense, science is part of a total transformative system of practices. Science constitutes an important aspect of the "knowledge" half of spiritual progress (the other "half" being righteousness, or ethical and physical practices).[20]

The transformative process also illumines the trajectory of ethics, as noted in Plato and Kierkegaard, above. Here, the basic principles of social and religious ethics are conventional in nature (at best), and the "spiritual" stage of progress (enlightenment) leaves those conditions behind. However, in certain spiritual traditions of early Christianity, the transcendence of ethical claims and categories of enlightenment is not the end of the process. Rather, there is a further stage of reintegration (of subjectivity) in which the ineffable experience of the Divine is expressed as fully as possible in the saint.

Spiritual practices especially point to a curious sort of feedback loop, in which the attainment of the goal of a practice produces transformations in the practitioner that involve a further transformation of the practice itself. The entire practice, as it were, is shifted to a "higher" (or more potentiated) "level" of experience.

In regard to the system of spiritual practices as a whole, the entire process consists of practices of transformation: they lead to the reintegration of a fragmented self. But the perfection of reintegration continues to transform the self until it becomes utterly other than itself—utterly open to God and others. It then engages in a reciprocal transformation with God and others, becoming a perfect glass through which the transforming rays of God's energies can shine on and transform others, and all of natural being.

II. SCIENCE AS A TRANSFORMATIVE SYSTEM

A. Re-envisioning Science

Recent theories in the philosophy and history of science, and new approaches in the debate over the relationship (if any) between science and religion point to the need for a more general level of analysis. In the first place, the science and religion debate has produced little light, but much heat. I think it is safe to call the relation between science and religion a *differend*, in the sense given this term by Jean-François Lyotard. A differend occurs when two phrase regimes cannot be mediated—there are no common terms or shared concepts, and therefore communication between them becomes impossible. Defenders of science do not share a common conceptuality—in fact, in a sense they do not share a common experience of the world.

The complexity thesis is one recent attempt to overcome this differend. It reminds us that any generalization we make about science or religion risks ignoring the historical, social, institutional, and ideological differences embodied in those concepts. "Science," for example, can connote any organized form of knowledge, the totality of the sciences, a common methodological approach (scientific method—a controversial concept itself), or the valorization of a certain kind of knowledge. Parallel complexities, of course, are embedded in the concept of religion.

The complexity thesis provides an important reminder not to reduce or reify our categories. However, often the presentations of complexity are rhetorical devices used to defend religion by pointing out the historical interactions of science with religion. Science emerged, the argument goes, from a religious milieu; most of the early practitioners of science were religious people; science and religion share some ontological and methodological similarities. This argument ignores the contingency of historical claims—the fact that it is notoriously dangerous to base claims about the nature of something on contingent historical facts. In addition, there is a suggestion of the genetic fallacy here: science, the story goes, has its historical roots in religion, therefore, science today is intrinsically religious in structure, in methodology, or in its very complexity—and in the last instance, the argument becomes viciously circular.

A further problem with the complexity thesis is that its analyses tend to remain at the level of the phenomena that the theories purport to describe. We see this problem most acutely in the antireductionist challenge to reductionist theories of science. The need to maintain different levels of analysis, and not to reduce one level of analysis to another—even if this means that risking a certain messiness in our analyses:

> Distinctive and independent concepts for each level remain entrenched, not merely as necessary for describing phenomena and their laws at different levels, but as integral components of the theories that remain independent of theories designed for levels above and below...The complete explanation of any phenomenon will require different accounts from all of its levels. This does not reflect the neatness and elegance of a reductive account...but antireductionists maintain that only this approach can account fully for the way things really are.[21]

The "systems-theoretic" approach of Evandro Agazzi provides a method of analysis that avoids these issues: the angsty skepticism of some versions of the complexity thesis and antireductionist approaches; as well the circularity and historical contingency of many the claims deriving from these approaches. Agazzi uses systems-theoretic to get below the surface complexities of science as a system, and to provide an astute analysis of the mechanics of such a system. His level of analysis is general enough that it can apply to various kinds of system, but not so general that it becomes a description merely of anything whatsoever that exhibits any kind of organization or systematicity.

In his work, *Scientific Objectivity and Its Contexts*, Agazzi argues, as the title implies, that science, at any particular time, in any particular sub-discipline, constitutes a seemingly necessary nexus of metaphysical assumptions, theoretical givens (received wisdom or well-attested findings), and methodological approaches to the given science. The various components of such a nexus, however, do not necessarily develop or change at the same rate. For example, Agazzi shows how the Aristotelian metaphysic that seeks essences persisted long after more "modern" notions of the relation of theory to observation, and of methodology, could have supplanted it:

> The criticism of Aristotelian physics, for example, was promoted within the Aristotelian framework itself, so as to improve or perhaps correct it, but not to reject it. This can be said because the fundamental points of view, the conceptual tools, the categories applied and, more particularly, the aims of natural investigations were the same as those of Aristotle.[22]

Agazzi's systems theory approach allows his to generalize from this historical development to the nature of a system itself:

> Under this respect, the scientific system shares one of the most characteristic properties of the social systems in general; that of being able to give very flexible answers to its environmental conditions, by modifying its internal structures, its ways of functioning, by redefining its goals in order to go on in its essential structural characters.[23]

Agazzi views the regaining of systemic equilibrium as one of the primary functions of a system, including the "scientific system." At the same time, he notes that the variables of a system form a "range," and that as long as these variables are maintained within a "critical range," the system can reproduce itself. In this light, systemic change is regarded as a challenge to the coherence and continuity of a system. The standpoint of transformativity reveals something different: that change is not a challence to a system, but is the lifeblood of a system.

Here, I think the Hesychastic model provides a key alternative insight. Hesychasm presents us with a model of a *transformative system*, one in which the very nature of the system is transformation. In the next section, we will examine a specific example of such a system, that of Gregory Palamas.

B. Science in the Hesychastic Theology of Gregory Palamas

It is important to note, first, that Gregory does not reject the value of philosophy,[24] or what we would call secular learning (including the natural sciences), in general. It is true that he refers to it as "demonic"—but only insofar as its "goal" is not knowledge of God (in which case, as we will see, it is not truly "science" in the fullest sense). Therefore, Gregory recontextualizes learning, encompassing all learning within the ultimate end of knowledge of God. For the same reason, it is inaccurate to claim that "Palamas denies secular studies are a *necessary prologue* to such deeper insight."[25] While Palamas *would* say such a thing if he had experience of our modern division of religion and science, he simply could not say such a thing in his day. He does not deny secular learning any relationship to divine knowledge; rather, he is at pains to maintain their *true* linkage:

> But if one says that philosophy, insofar as it is natural, is a gift of God, then one stays true, without contradiction, and without incurring the accusation that falls on those who abuse philosophy and pervert it to an unnatural end. Indeed they make their condemnation heavier by using God's gift in a way unpleasing to Him.[26]

The work opens with a question posed by someone, perhaps a younger monk, worried by the claims of pagan philosophers to know the causes and understand the movements of natural beings and forces.[27] By placing these concerns in the mouth of a pious monk, Gregory contrasts the pride of the pagan philosophers in their knowledge with an exemplar of true Christian humility, who admits that, when he first heard the philosophers' claims, he did not know how to respond to them.[28] That is, monastic authors seek not only to impart a teaching to the reader, but to instantiate or exemplify the very virtues that they discuss. In addition, the dialogue form deflects the authorial role, and places the burden of responsibility—not only for interpreting but also for living out the text—on the reader. Moreover, the function of dialogue is also to exhibit that the theological "system" is itself transformative, insofar as the reader engages it as a participant. In this sense, dialogue instantiates the very synergetic relationship between human spiritual practices and divine grace.[29]

Gregory's arguments are rooted not only in Christian tradition per se, but also in the specific praxis-oriented spiritual theology of Hesychasm, an integrated system of spiritual practices that transform human nature.[30] Gregory writes from the context of, and assumes the reader's familiarity with, this system of monastic spiritual practices.[31] Here, we will only consider some of these features that are relevant to the immediate purpose of understanding Gregory Palamas' take on the relationship of secular (scientific and philosophical) knowledge and Christian spirituality.

In the monastic, and more specifically Hesychastic, literature, the practices of transformation have a twofold character: they are, in different senses, both means and ends. For example, the practice of prayer may connote the different forms of prayer practices laid out by Origen in his *De Oratione*, such as thanksgiving, praise, petition, and intercession. These types of prayer all produce transformation of human subjectivity by different means. On the other hand, "prayer" in the monastic text may refer to the goal of prayer practices, namely "pure prayer," or a state in which prayer is continuous and undisturbed by appetites, emotions, passions, and thoughts. In addition, the practices also overlap, or are mutually interconnected. So, "pure prayer" is basically consonant with the practices of attentiveness (or watchfulness), purity of heart, and *Hesycheia* (tranquility).

Gregory's treatment of secular knowledge places that knowledge within the context of the monastic spiritual system. Thus, like the other practices, such "secular learning" connotes both a set of preliminary, transformative practices, and also the goal of those practices, and that these practices operate within a context of mutually informing practices. Gregory assumes, then, that secular learning is not divorced from theology or spirituality, but has its place within traditional monastic theology and practice. What is this place?

In the Greco-Roman schools of philosophy, philosophy, theology, and the physical sciences were not yet divided. As Pierre Hadot has reminded us, even the study of *phusis* was not a merely rational or experimental interpretation of physical reality, but was what he called "theological physics."[32] Physics, in this sense, was the study of the cosmological and providential plan that gave meaning and place to all things that exist. Without knowledge of this plan, humans cannot be happy, cannot achieve their *telos*. Thus, the goal of physics, no less than the goal of philosophy, was to provide meaning and happiness to human beings.

As such, the study of natural causes was itself a contemplative practice. That is, the goal of this study was to contemplate the providential arrangement of natural things, and the place of human beings within it. This means that the study of natural causes formed part of the intellectual side of the system of spiritual practices. In traditional spiritual practices, this intellectual side is complemented by an ethical side, consisting of practices of moral purification and upbuilding. Later, I will address the specific problem raised by this twofold structure of spiritual progress.

This approach to the study of nature was carried over into early Christian theology. As in Hellenistic philosophy, the study of natural causes, and secular learning more generally, is not irrelevant to spiritual progress, but has a very specific role within the system of practices that make that progress possible. What is most significant about Gregory's argument is that he does not conclude that pagan learning is useless, but only that it is dangerous when pursued outside the context of the faith; outside, that is, the system of transformative practices. Gregory's position regarding pagan philosophy is a specific instance of the more general principle that he cites: "It should be remembered that no evil thing is evil insofar as it exists, but insofar as it is turned aside from the activity appropriate to it, and thus from the end assigned to this activity."[33]

This means that scientific knowledge, specifically, and all knowledge generally, has an intrinsic end. This, of course, is an idea dating back at least to Aristotle (e.g., the opening of his *Nicomachean Ethics*). But Gregory ties it to the Christian doctrine of the image of God. Gregory argues that it is not knowledge *per se* that is the image of God in man, but the correct use of knowledge, the ability to "divide" correctly between the false and true.[34] So, the goal of knowledge, intrinsic in knowledge itself, is discernment.

Discernment is a complex virtue in the patristic tradition. First, occurring as it does toward the end of the purification stage of spiritual progress, discernment is the culmination of all the prior practices of purification.[35] In fact, discernment shares many characteristics with the charismata, or spiritual gifts that properly belong to the second stage of spiritual progress, enlightenment. Second, discernment is also complex in the sense that it refers to a variety of practices that may seem at first glance unrelated. Discernment connotes the discrimination of the spirits, the ability to identify good and evil spirits (angels and demons). Associated with this ability are practices such as rejecting angelic apparitions, in case they might be demonic deceptions. This practice assumes that, at the beginning of the spiritual "ladder," our ability to tell the difference between angels and demons will be meager, and our desire for spiritual experiences (rather than *Hesycheia*) will lead us to court such experiences, to our peril. Typical of many Hesychastic practices, however, discernment also refers to more advanced practices, such as discerning the state of the soul of another person, a type of discernment that we find in enlightened elders.

All of the practices of discernment, however, have in common the spiritual necessity of "dividing" between the true and false. As such, they all share a common spiritual prerequisite, namely dispassion. Stoic and Epicurean philosophers had debated the exact nature of dispassion, and whether it was a desirable or even attainable end, and there were differences of opinion on the matter even amongst Christians. The basic point of contention was whether passions should or could be eradicated entirely, or whether they were instead to be controlled; and whether, in the latter case, they had some positive roll to fulfill. As Paul Blowers has shown, the point was a contested one from at least the time of Plato, and through many of the pagan Hellenistic schools. As Blowers mentions, the early Christian theologians/monastics continued the debate over this topic. Most of the Christian monastic writers held the position that the passions must be controlled, or better directed, but that they were an intrinsic, or even essential, aspect of the total human being.[36]

Gregory regards dispassion as the prerequisite for any profitable exploration of natural being:

> A life which hope in God has liberated from every care naturally impels the soul towards the contemplation of God's creatures. Then it is struck with admiration, deepens its nderstanding, persists in the glorification of the Creator, and through this sense of wonder is led forward to what is greater."[37]

To get at what Gregory is saying, perhaps we should ask what the alternative approach he is condemning consists of. It would be an approach dominated by "care" and pride,[38] that it operates outside faith (a phrase that needs unpacked), and that it operates outside the end appropriate to it. While these phrases are somewhat opaque, we can see that they each and all apply well to the way in which modern science works, so much so that we may assume that Gregory was addressing something very like modern science. Here, the approach is dominated by care and pride—that is, a concern that nature is dangerous, and needs to be controlled, and a belief that human beings should be in control of it, and the goal of scientific knowledge is such control. In order to do so, both the science of Gregory's time as well and modern science have argued that faith will restrict the pursuit of such control. However, if in fact science should not be about control, but about providence—that is, finding our place within the created order, cooperating in the process of spiritually transforming that order according to God's will—then modern science has distorted its own end.

Gregory opposes the care, pride, and lack of *telos* of science to the benefits of revelation: "Do you not clearly see that it is not the study of profane sciences which brings salvation, which purifies the cognitive faculty of the soul, and conforms it to the divine Archetype?" So, in order to use scientific reasoning well, we should be saved (an assertion that needs unpacking, but at the least connotes baptism, also referred to as "illumination" in the East), purified (i.e., the practices of the first stage of spiritual progress), and transformed in our *nous*—in other words, enlightened (transcendence of the subject/object distinction, the law of non-contradiction, the categories, etc.). So, in order to use scientific reasoning well, we must achieve the goals of the first two stages of spiritual progress.

Interestingly, this means that we can only pursue the true goal of science if we transcend or transform the ways in which science is usually done, in fact, only if we transcend the very conditions of rational knowledge:

> But what has already been said should suffice to demonstrate easily to the unconvinced that there is indeed an intellectual illumination, visible to those

> whose hearts have been purified, and utterly different from knowledge, *though productive of it.*[39]

This, then, raises the question of the connection between illumination and knowledge. Illumination requires purification, and this includes all the practices and virtues that we have examined, above. These practices and virtues, in turn, prepare the individual practitioner for the transcending of natural knowledge, the transformation of their natural being. This involves the transcendence of the intentional structure of consciousness, the law of non-contradiction, the very categories of consciousness. These practices and virtues make possible the contemplation of created essences.

Gregory gives us an example of this type of contemplation in the first half of his *150 Chapters*. Here, in the first place, Gregory clearly delineates between natural knowledge and spiritual knowledge, and demonstrates their proper relationship. As Doru Costache comments, "The message conveyed is transparent: on the one hand, there are areas of confluence between theological and natural epistemologies; on the other hand, there are domains that cannot be dealt with outside the confines of divine revelation."[40]

Gregory provides us with a naturalistic (as opposed to spiritual) account of the beginning and end of the world (cosmology), an account based on the testimony of nature and history, corrected according to the teachings of Moses, and utilizing the logical argument known today as the cosmological proof for the existence of God.[41] He then raises numerous questions concerning the teaching of "the ancient Greek sages" concerning the world soul.[42] Here, he gives a natural explanation for the very phenomena for which the Greek sages had adduced the "world soul." In the same way, Gregory also rejects the Greek sages' teaching on the habitable regions of the earth.[43]

Having provided this account of the natural world—a naturalistic cosmology—Gregory goes on to consider the nature of our knowledge—a naturalistic epistemology (which implies a broader anthropology). Gregory's epistemology highlights the difference between natural knowledge and supernatural (spiritual) knowledge. Natural knowledge consists of the joining (the Fathers used the term "coupling") of a sensory stimulation with a passion. In the language of Maximus the Confessor, this "coupling" of image and passion constituted the beginning of an almost—but not necessarily—automatic process of sinning.[44]

In contrast, Gregory posits a type of intellection free of any sensual input—an intellection that in fact transcends all the conditions of natural knowledge. This corresponds in the Hesychastic tradition with the third stage of spiritual progress, illumination (or enlightenment). While Gregory does not spell out most of the details of illumined knowledge here, he does mention several of them, and we can fill in the details from other Hesychastic texts.

Gregory begins by establishing the conditions of natural knowledge, in order to show the contrast with supernatural knowledge. In the first place, natural knowledge consists of inference from sensory experiences,[45] "things of the Spirit being beyond its scope."[46] There are three important issues here. First, inference is not infallible. Any knowledge based on inference may be overturned by further evidence. Of course, this uncertainty is the glory of modern science, but Gregory regards this uncertainty in contrast to unitive, enlightened knowledge, which is unconditioned and certain, and based on direct experience.

The second problem with natural knowledge, as I mentioned above, is that intellectual images of sensory experiences may stimulate the passions, resulting in a sinful attachment to those objects and experiences. In the context of his discussion about the role of natural knowledge, and especially scientific knowledge, Gregory's concern is not simply the danger of the passions at an individual level, but the passionate nature of natural knowledge as a whole. Gregory regards this feature of natural knowledge as the root of the vanity of natural knowledge and the egotism of its exponents. A key point here is that Gregory does not advocate the extermination of the senses or the passions, but their transformation, their deification. This explains the role of the "spiritual senses," a traditional doctrine that Gregory also appeals to, in which humans are said to possess a set of senses for perceiving spiritual realities (including creates essences, angels and demons, divine providence, and miracles). While everyone possesses the spiritual senses, the natural senses distract us from them, and tend to suppress them. The transcendence that coincides with enlightenment frees the practitioner from enslavement to the senses and passions, and allows for the possibility of their transformation.

Another condition of natural knowledge is the separation of knower and known. This means that natural knowledge cannot attain to enlightenment, much less to the unitive experience of God. By its nature, natural knowledge is cut off from communion with the objects of its study. This is perhaps the most tragic consequence of the Fall, that human nature now prioritizes the desire for a kind of knowledge that can never be satisfied. In addition, natural objects distract us from knowledge of our self, involving us in a self-perpetuating illusion. As I will discuss below, Gregory believes that enlightened self-knowledge gives us the ability to discern the causes of natural objects, that is, their place within divine providence.

Gregory mentions other conditions that pertain to natural knowledge, such as its limitation to conditions of space and time and its reliance on opposites. In both cases, the unitive "experience" of God transcends these conditions. For all of these reasons, because of all of these limitations and dangers, Gregory describes natural knowledge as the plaything of the devil.[47]

Having sketched the contours of natural knowledge, Gregory turns to spiritual knowledge, which is nothing less than knowledge of God, the world as a whole, and ourselves. This knowledge is spiritual in the literal sense—that is, it comes from the Holy Spirit:

> The human soul is something great and wondrous, superior to the entire world; it overlooks the universe and has all things in its care; it is capable of knowing and receiving God, and more than anything else it has the capacity of manifesting the sublime magnificence of the Master-Craftsman. Not only capable of receiving God and His grace through ascetic struggle, it is also ableto be united with Him in a single hypostasis.[48]

Here we have most of the characteristics of enlightened knowledge: it transcends worldly conditions, including space and time; it knows God through union with God; it understands the meaning of natural causes, it knows their roles within providence, and it can therefore care for them properly.

His contrast between natural and enlightened knowledge leads Gregory into a paradox[49] : if true knowledge transcends the conditions of natural knowledge, how does it relate to or effect natural knowledge? To put it another way—and to move with Gregory from the consideration of the merits of "pagan philosophy" to the consideration of the role of *apophasis*—the true pursuit of knowledge of created being depends upon the proper understanding of the relationship of *cataphasis* (positive predication) and *apophasis* (negative predication). To put it simply, the latter (transcendence of the canons of natural knowledge) is the condition for the proper exercise of the former (secular knowledge, science):

> This is why every believer has to separate off God from all His creatures, for the cessation of all intellectual activity and the resulting union with the light from on high is an experience and a divinising end, granted solely to those who have purified their hearts and received grace.[50]

The transcendence of the canons of natural knowledge[51] is not a mere process of abstraction (such as Gregory attributes to the Messalians), nor a mere linguistic negation; rather, it is "accomplished not only in words, but in reality."[52] Moreover, as the phrase "purified their hearts and received grace" indicates, this true knowledge is the product of the synergy between human efforts at purification (ethics) and divine grace.[53]

Recent treatments of "apophatic theology" have tended to ignore this point. Too often, they reduce *apophasis* to a linguistic technique of negation. In fact, the very characterization of this "way" as "apophatic" or "negative theology" is misleading. What characterizes this "way" is not, in fact, practices of negation, but the goal of these practices, namely union with God, or what Gregory usually characterizes as "purity of heart." This purity of heart is not mere absence of all characteristics, however:

> There are other things, and many of them: There is the pledge of things promised in this life, and also the blessings of the life to come, which are rendered visible and accessible by this purity of heart. Thus, beyond prayer, there is the ineffable vision, and ecstasy in the vision, and the hidden mysteries.[54]

Purity of heart, then, means the opening of the practitioner to the fullness of reality, not its rejection. Purity of heart overcomes the hubristic attitude, not only of scientific knowledge, but of all human endeavors. Purity of heart makes the transformation of the self perpetual, and makes the individual the agent of transformation in others, in society, and in nature.

Equally important, the apophatic does not merely negate the cataphatic, as if the principles of created things are unimportant. On the one hand, "natural" knowledge does lead "up" to purity of heart. On the other hand, however, purity of heart is the fulfillment of those principles:

> For, according to the teaching of the Fathers, every divine command and every sacred law has as its final limit purity of heart; every mode and aspect of prayer reaches its term in pure prayer; and every concept which strives from below towards the One Who transcends all and is separated from all comes to a halt once detached from all created beings.[55]

Purity of heart, then, is not "empty" of "content," but this "content," Gregory emphasizes, is "ineffable," and can only be represented in symbols and signs. Therefore, to interpret these symbols correctly, it is necessary to have attained the same purity of heart that they represent. Here, in addition to attacking Barlaam the Calabrian and others who accused the Hesychasts of claiming to see God with their natural senses, Gregory is also establishing a fundamental point about the relationship of "natural" knowledge and "supernatural" knowledge. The latter requires a prior transcendence of the natural canons of knowledge, which is the product of the transformations of human nature by the cooperation of the practices of purification and God's self-revelation (the divine energies, or grace).[56]

Purification is necessary, not because the body is evil, but because the "mind" has become subject to the passions.[57] We must free ourselves from our tendency toward "dwelling on fleshly thoughts," in order to reestablish "the oversight of the mind, and in this way establishing a law appropriate for each power of the soul, and for every member of the body."[58] Indeed, the virtues are nothing less than the proper ordering of the faculties. The proper ordering of the senses is temperance, that of the emotions (the affective) is love, and that of the mind is watchfulness. Here, Gregory gives a veritable précis of the Hesychastic stage of purification. The senses are controlled by utilizing practices of temperance, i.e., practices that disallow the appetites from overstepping their natural boundaries. These practices would include fasting, limiting sleep, chastity, and vigils. The emotions are controlled, not by cutting off "wishes and desires," but by using them as an opportunity for righteousness. These practices—in Hesychastic literature often simply called "the practice of the virtues"—include cutting off the sequence of sin, avoiding situations that contribute to sin, and actively practicing the virtues. The mind attains watchfulness through prayer, by means of which it "sees" that grace imparted to it by God.

Gregory makes a critically important point—both for our understanding of Hesychastic spirituality and for our understanding of the relationship of this spirituality to scientific knowledge—and that is that this vision of the grace given to us occurs *within* the body. "Can you not see, then, how essential it is that those who have determined to pay attention to themselves in inner quiet should gather together the mind and enclose it in the body, and especially in that "body" most interior to the body, which we call the heart?"[59]

As Gregory insists repeatedly, *apophasis* is not mere negation, or dialectic, which are possible by natural means. Rather, true *apophasis* leads to transcendence of all predication. This experience is not "empty," however. Rather, what is found with the cessation of thought and the experiences of the lower faculties (passions, emotions, appetites) is the "dark light" of God's presence within us.

The goal of the Hesychastic practice of "bringing the mind into the heart" is to retain this presence within the body. Hesychasm does not lead to an "out-of-body," mystical experience, but rather seeks, through the agency of the Holy Spirit, to transform the body—and all of human nature—from *within*.

> The spiritual joy which comes from the mind into the body is in no way corrupted by the communion with the body, but transforms the body and makes it spiritual, because it then rejects all the evil appetites of the body; it no longer drags the soul downwards, but is elevated together with it. Thus it is that the whole man becomes spirit, as it is written: 'He who is born of Spirit, is spirit.' All these things, indeed, become clear by experience.[60]

This is the experience behind the Hesychastic employment of the image of the Light of Mount Tabor,[61] as well as behind the famous scene (in Byzantine Christian circles) in which Seraphim of Sarov appears to his disciple, Motovilov, as bathed in light.[62] In the latter instance, we also have the beginning of a sense of how the transfiguration of human nature of an individual can lead to the transfiguration of others, and potentially, of all of nature. When Seraphim asks Motovilov why he averts his eyes, Motovilov replies that Seraphim's face is too bright to look at. To this, Seraphim replies that, if he only knew it, Motovilov's own face is equally bright. Seraphim generalizes this experience in his most famous apophthegm, "Acquire the presence of God, and thousands around you will be saved."

While this suggest the connection between enlightened nature and "natural" knowledge, it still leaves the problem of explaining how the transcendence of the canons of natural knowledge can be "translated" into everyday experience. To put the problem more precisely, we can examine the issue of ethics as an example of the broader problem. In the first place, the practice of the virtues is central to the first stage of spiritual progress, namely purification. It consists of practices to foster ethical behavior, and to overcome the habits, and resist the temptation, of evil. Thus, the practice of the virtues relies on the firm distinction of right and wrong, good and evil, virtue and sin. However, at the stage of enlightenment, the aspirant transcends all natural ethics, including the very law of non-contradiction that makes the distinction of good and evil possible. If we achieve this stage of spiritual progress, then, how can we be sure that this enlightened knowledge will lead to ethical actions? In fact, there are all-too-many examples of persons who have attained enlightenment who have used this knowledge to control others, to commit evil acts, and to make the Faustian bargain. How, then, can we be sure that our progress toward enlightenment will result not only in increased knowledge and power, but in righteousness?

This problem is fundamental in the works of Gregory Palamas, Maximus the Confessor, and others. The system of spiritual practices of Hesychasm always operates at two distinct, but related, levels: knowledge and righteousness. The relationship between them is complex. On the one hand, righteousness is a condition for the attainment of enlightenment. On the other hand, enlightened knowledge transcends all conditions—and therefore can transcend righteousness itself. This paradox is rooted in a fundamental experience of the spiritual life, as we already see in the lives of the Desert Fathers and Mothers. Here we find many admonitory tales of monks whose wisdom has outstripped their righteousness: the monk convinced by demons that he could fly, and so he dies jumping down a well. Gregory seems to regard the issue much more generally, however, as the danger of a kind of knowledge—and culture—that serves the passions. In this regard, today we might have examples of the dangers of which Gregory warns in the case of spiritual teachers and gurus who (perhaps) have attained enlightenment, but who are driven by their passions (and, Gregory would say, demons) and therefore lead others astray.

Jacques Derrida coined the term "hostipitality" (combining "hospitality," a key Biblical and monastic virtue, and "hostility") to connote an unconditioned openness to the radically Other (which is what enlightenment attains).[63] Such an unconditioned form of knowledge would seem to render the practitioner subject not only to angels, but to demons; not only to Messiahs, but to monsters. The Hesychastic tradition would argue that Derrida misses the role of the cataphatic in relationship to the apophatic, the relationship between natural knowledge and enlightened knowledge. Freedom from the passions (through purification) enables the *nous* to transcend the conditions of natural existence, and this in turn makes possible a direct perception of natural causes. This perception of natural causes does not stimulate the passions, and therefore the increase in knowledge does not produce demonic delusion. To use the example of monastic hospitality, one receives the other without condition, as hospitality demands, but central to true hospitality is *faithful* preparation for that other. Hesychastic practice consists of nothing less than the preparation (conditions of knowledge) for the radically unknowable (enlightened knowledge), that is, discernment and faith.

III. IMAGINING TRANSFORMATIVITY

Gregory Palamas' criticisms of Byzantine culture—of the role of science, and pagan learning generally—rely, at least in part, on *a priori* reasoning.[64] In its most obvious form, such reasoning simply takes scripture as authoritative (e.g., in regard to the geocentric nature of the solar system), and rejects any scientific findings that contradict revealed witness. Sometimes, the reasoning is less clearly *a priori*. In some cases, the scripture is not directly appealed to, but what is appealed to in its place is a cosmology implied by the scripture. On the basis of this cosmology, various kinds of things may be inferred, such as conclusions regarding the distribution of the lands of the earth and other natural facts and principles; ethical and religious principles; and a wealth of metaphoric meanings. Gregory's theology displays a rigorous consistency across these various kinds of knowledge; that is, within his Christian symbol-world, all the pieces fit. However, such consistency is not the same thing as scientific proof. In fact, it is precisely Gregory's style of *a priori* reasoning that modern science has tried to overcome through scientific method, experimentation, and mathematics. Nevertheless, this type of reasoning has a significant historical importance; comparatively, the method of modern science is very much an historical latecomer. More important, *a priori* types of reasoning still remain important in almost all areas of culture, from political speeches to sermons. Nor is this a style of reasoning particular to any party affiliation, religion, or other distinction. It is a disease—if a disease it is—shared by all (even scientists).

We can see how hard it is to overcome the *a priori* tendency in the struggle of the early Enlightenment scientists to develop the modern scientific method. From at least the time of Roger Bacon, scientists were busy expanding not only our knowledge of nature, but also formulating a method that would yield knowledge. In this process, we see the very meaning of knowledge itself shifting, subtly but radically over time. *A priori* reasoning long maintained its purchase in the work of these early modern scientists. Of course, the dominant form of scientific method taught and used today finally overcame the most egregious forms of *a priori* reasoning. Nevertheless, the *a priori* method has continued to have at least a cultural influence within various forms of non-mainstream science (such as Goethe's theory of optics), occult religion and theosophy, and various kinds of conspiracy theory.

So the question is: if this kind of reasoning is so important to human culture and history, what is its relationship to scientific reasoning? Most scientists[65] would probably argue that scientific reasoning produces revisable facts, while *a priori* reasoning tends to be dogmatic and therefore not revisable. Since the conclusions of *a priori* reasoning are often not testable (and at any rate are not revisable even if they are testable), they tend to be private: the dogma or ideology of a particular group, race, religion, or political party. For these reasons, *a priori* reasoning is the basis of the type of logic known as "conspiracy theory." Here, mistrust forms a kind of cosmology, so that within this cosmology, everything is meaningful (as for Gregory's Christian cosmology, as well), but also the meaning of anything that happens *must* imply other meanings: such as a government cover-up, a second shooter, or some other nefarious purpose or practice. Here again, the connections between cosmology and inferences are logical, though they are often of the sort that Umberto Eco attributes to the "moron." The "moron" is a "master of paralogism": "in such statements you suspect that something's wrong, but it takes work to show what and why."[66] This is not to say that all forms of *a priori* reasoning are moronic, or conspiratorial, but certainly *a priori* reasoning underlies these styles of thought. The question is whether there are any forms of *a priori* reasoning that are not moronic or conspiratorial, that even if not scientific (in the modern sense) are justifiable, that produce *knowledge*.

The immediate problem is that "knowledge" can mean so many different things. To ask a Socratic question: how can we know whether *a priori* reasoning is knowledge, if we don't first know what knowledge is? And if we don't know what knowledge is, how shall we search for it, since we lack a standard for knowing what it is, even if we come across it?

All forms of knowledge—all types of natural knowledge, and enlightened knowledge as well—comprise a cosmology (*logoi*, or the essences of created things); an anthropology (i.e., a conception of the structure of human being, e.g., the tripartite self); and an ethic (practices that enact the cosmology and anthropology). The relationship between these is one of mutual reference: the cosmology, anthropology, and ethic are mutually reinforcing. I will refer to this relationship as "practice." Practice, in this sense, resembles to some degree what William Paden has called "religious worlds."[67] Paden describes a religious world as:

> the comparative category par excellence. All religions inhabit worlds constructed by their own particular religious symbols. Indeed, what the world 'is' has traditionally been defined by religious language, though now science has taken over the task for many. *World* here is a descriptive word for what a community or individual deems is the 'reality' it inhabits, not a term for some single system objectively 'out there' that we all somehow share. The guiding principle of comparative study must be that each religious community acts within the premises of its own universe, its own logic, its own answers to its own questions.[68]

Paden defines a religious world primarily in terms of the external symbol system that expresses it. However, in order to incorporate not only natural knowledge, but enlightened knowledge, in our definition of "knowledge," we must expand the concept of world beyond Paden's definition. Here, I believe that Mark Peckler's concept of the "transcendental power of imagination" is crucial.[69] Peckler's definition of this concept relies on a Heideggerian reading of Kant (or a reading of Kant within the horizon of Heideggerian concerns). The transcendental power of imagination is not fantasy, though the ability of the imagination "to bring to image or 'make present' that which is not or cannot be present-to-hand 'empirically' is at least in part responsible for such common views of the imaginal that stress this unreality."[70] Rather, transcendental imagination is "the ontological horizon of encounter with beings," and "as horizon of inner sense and the pure image of the schemata of concepts," it "does not create beings, but it does create the ontological horizon of encounter with beings."[71]

In this sense of imagination, a practice (such as systems of spiritual practices, science, and other structures in which cosmology, anthropology, and ethics mutually transform) consists not only of theoretical (intellectual, noetic) and practical (ethical, purificatory) concerns and practices, but is the conditionless condition that forms the horizon of possibility for such concerns and practices. In my own view, this is a place that is not a place in which the mutual and continuous transformations between conditions and formlessness, between natural knowledge and supernatural (enlightened) knowledge, between cataphasis and apophasis are perfectible (teleological and potentially progressive).

By this standard of knowledge, modern science is as much a form of *a priori* reasoning as Gregory's theology. Each system of thought is a relationship between a cosmology, an anthropology, and an ethics—a "practice." Each practice is rooted in the pre-cognitive horizon of imagination. This is not to argue that somehow, because of this common "source" in the imagination, all symbol systems are equal.[72] But they are all "rooted" in a "source" that is "beyond" them, providing them the very possibility of possibility.[73] How this conditionless state of possibility instantiates in images depends upon what conditions (cosmology, anthropology, and ethics or "practice") are chosen. This in turn will determine the practice that forms. The conditions chosen will also be the conditions that limit or free the practice, in terms of transformation.

While each world is self-referential, it is also incomplete—the coherence of its cosmology/anthropology/ethics (=practice) is based on an assumption (or assumptive reality or being) that transcends and cannot be reduced to that world itself. This is the transcendental imagination. All practices (spiritual practices, science, or others) share this same transcendent origin.

But if the "origin" of knowledge transcends the conditions of knowledge, then how does this argument apply to science, for which the world is strictly immanent? Perhaps the account would be something like this: the immanence of the world forms the cosmological system of—the largest imaginal context for—the practice of science. Scientific method, experimentation, and mathematization, along with the ideology that supports these (the ideals of objectivity and elegance, the public nature of the scientific enterprise, the always revisable nature of theories, etc.), comprise the ethic that supports and is supported by this immanent cosmology. These, in turn, produce and are reproduced by a certain "subjectivation" (to use Foucault's term) or anthropology. The scientist is a rationalist (perhaps an atheist or at least agnostic), committed to public and honest research, always trying to remain objective, always willing to submit findings to further research and experimentation.

While this seems an entirely closed system, it is not. Certainly, unlike Gregory's theology, it does not overtly refer outside of itself to a transcendent Source. Indeed, modern science seems utterly without transcendent basis. In modern science, the cosmology is utterly naturalistic, as are the anthropology and the ethic. But the apparent absence of transcendence points us to the particular kind of transcendence of science. Science assumes that there is one kind of knowledge, one kind of object of knowledge (matter), that there are specific methods that can achieve that knowledge, and one kind of person (the objective, unemotional scientist) who can achieve it. These assumptions are mutually implicative. The immanence of science therefore relies on an unproven coherence of assumptions that science asserts dogmatically as the only means of obtaining knowledge. The revisable nature of scientific knowledge does not lessen the dogmatic nature of these assumptions. These assumptions are nothing but the symbolization of the pre-symbolic imagination peculiar to science. It is self-manifesting, and self-affirming, like all practices. The question is, on the basis of the single valid criteria for judging between symbolizations: does it transform, and in transforming, is it transformed?

Herbert Marcuse, in his *One-Dimensional Man*,[74] argued against certain types of analytic philosophy that they not only lack any critical edge, in fact, they acted as the ideological justification of the *status quo*. They acted, that is, according to Marx' definition of "ideology," and Marcuse thought that this was a betrayal of the critical spirit of philosophy. Many recent epistemologies seem to fulfill a similar function. In regard to *a priori* knowledge, the epistemological theories focus on the issue of justification. This issue is fraught with contention, but all the participants in the debate seem to accept without question that justification is a purely rational problem. But the question of whether *a priori* reasoning is ever justified cannot be answered without reference to the cosmological, anthropological, and ethical assumptions of the act of justification itself.[75]

Returning to the question of whether all practices are equally valid, now that we have established a fuller definition of these terms, and have argued that science also is such a world or practice, we can begin to answer this question. The hallmark of a practice is the mutually transforming nature of its cosmology, anthropology, and ethic. There is no universal standard to judge between worlds, except the extent to which they are transformative. We certainly can judge them on the basis of the products of their transformations, but this too requires a standard of judgment outside the world itself. Rather, the degree to which the world transforms and is transformed provides the only context-free measure that we can use to adjudicate between worlds. On this basis, modern science transforms, and is transformed by, the natural world, but its transformations are limited to its assumptive immanence. We see this in the technologization of society—the tendency of science to ignore causes and address problems as symptoms to be eliminated by the application of a better technology. We see the limitations of modern science also in the ideology that supports this technologization, an ideology as we have seen that is little more than a justification of immanence itself. We especially see the dangers of such immanence in the books and movies that depict a not-too-distant future society in which science has eliminated all human values and natural being.

A system like Gregory's Hesychasm, on the other hand, produces transformations that have as their source and goal an infinite God, and therefore these transformations are endless and perfecting. Within this world, science itself finds its true context and meaning, as the contemplation of natural causes for the purposes of deepening our relationship to God. Science itself transforms nature (including human nature), and thereby transforms itself, in a manner that endlessly reinforces immanence. This results in all the limitations of natural knowledge that Gregory notes (see above). Religious practice, on the other hand, transforms nature infinitely, in synergistic relationship with the divine.[76]

The theory of practice offered here provides the possibility of avoiding getting lost in metaphysical arguments that are not grounded in practice, of escaping the Scylla and Charybdis of reified dualism, and of producing yet another apologetic for science *or* religion. Science as a system cannot, by its own nature, and by the nature of social systems themselves, exist independently of other systems. Nor can it accomplish its intrinsic *telos* of natural knowledge without first transcending the limitations of natural knowledge. These two claims, I have argued, demonstrate the need for a deeper, more general analysis of the dynamics of systems of practice. By means of this analysis, we will find ourselves empowered to engage these systems of practice for the purpose of transformativity: the reciprocal and ongoing transformations of our self, others, and nature itself.

TRANSFORMATIVE CONCLUSION

In the epigraph at the head of this essay, Paul calls for the nature of his audience members to be so transformed that they can know the nature of nature itself. However, in this very compact statement, Paul goes beyond the processes of enlightenment to the practices of perfection, or *theosis* (divinization). Purification of the person leads to knowledge of the *logoi* (essences or causes) of creation. This knowledge, in turn, seems to provide the condition for the love of Christ—but note that Paul mentions no conditional relation between knowledge of the *logoi* and the love of Christ. This is because the love of Christ "passes knowledge," and it is only by surpassing the natural conditions of knowledge, however exalted, that we can know the "fullness of God." May it be so.

ADDENDA

Addendum #1: The Tripartite Self

The tripartite self is traditionally attributed to Pythagoras, but was first systematized by Plato. A "dipartite" version is also commonly found. Another traditional anthropology distinguishes the various faculties of body, soul, and spirit (the intellectual, the incensive, and the concupiscent).

Spirit Nous (the light of the mind)
    Rational thought
Soul Thinking (brain activity; "monkey mind")
    Will
    Passions (habitual emotions)
    Emotions
BodyAppetites
    Instincts
    Sensation

Addendum #2: Outline of the Hesychastic System of Spiritual Practice

A. Purification
  1. Negative Praxis—external
    a. Exile
    b. Solitude and Community
    c. Ascesis (Fasting, vigils, control of sleep, sexual desire, etc.)
  2. Negative Praxis—internal
    a. Obedience
  3. Positive Praxis (Upbuilding)
    a. The Practice of the Virtues
    b. Discernment
    c. Attentiveness
    d. Dispassion
B. Illumination
  1. Practices of Transformation
    a. The Spiritual Senses
    b. Charismata
    c. Perception of Created Essences
    d. Realization of Universal Selfhood
  2. Cataphatic Theology
C. Perfection
  1. Practices of Union
    a. Transcendence of the Canons of Rationality
    b. Perception of Providence
    c. The Vision of the Trinity
    d. The Nature of Union
  2. Apophatic Theology
    a. the "Hyper-"

Addendum #3: The Characteristics of Enlightenment

Perfect discernment—note that discernment is perfected at the point when the conditions for discernment are transcended.

Charismata; these are primarily spiritual experiences that have a bodily effect: perception of the light of Mt. Tabor; warmth in the heart; synaesthesia; levitation, telepathy, bi-location, etc.

Contemplation of the light of one's own mind
  -the nature of the observer self
  -self or no-self? -it depends on where you are
Spiritual hermeneutics

-The spiritual level of interpretation becomes literal—as in Dante, Paradiso 43–5, "There we shall witness what we hold in faith,/not told by reason, but self-evident;/as men perceive an axiom here on earth."

Transcendence of the Canons of Reason:

-Transcendence of subject/object (intentional) structure of knowledge—also known as unitive knowledge.

-Transcendence of laws of logic: the law of identity; the law of non-contradiction.

-Transcendence of the *a priori* structures of consciousness (space/time)

Transcendence of Ethical Law

-e.g., SK's teleological suspension of the ethical

Cataphasis and apophasis

-enlightenment marks the *telos* of cataphasis, and the point at which apophasis begins

-apophasis coincides (or consists of) the transcendence of the canons of reason, and the other characteristics of enlightenment listed here.

Cataphatic practices (the culmination of the cataphatic ascent):

Contemplation of created essences
Contemplation of Hierarchies
Contemplation of Providence
Contemplation of the System of Practice Itself
Contemplation of the Holy Trinity
Pure prayer
Perfection of the virtues

BIBLIOGRAPHY

Agazzi, Evandro. "A Systems-Theoretic Approach to the Problem of the Responsibility of Science." *Zeitschrift für allgemeine Wissenschaftstheorie/ Journal for General Philosophy of Science*, 18, (1987), 30–49

———. *Scientific Objectivity and Its Contexts*. Cham: Springer International Publishing, 2014.

———. *"How Can the Problems of an Ethical Judgment on Science and Technology be Correctly Approached?,"* International Journal of Technoethics, 1(2), 10–18, April-June 2010

———. "Systems Theory and the Problem of Reductionism." *Erkenntnis* 12 (1978), 339–358.

Agazzi, Evandro and Jeanne Ferguson. "The Historical Dimensions of Science and Its Philosophy." Diogenes 33 (1985), 60–79.

Alai, Mario and Gino Tarozzi, eds. *Science between Truth and Ethical Responsibility: Evandro Agazzi in the Contemporary Scientific and Philosophical Debate*. New York: Springer International Publishing, 2015.

Andersen, Hanne, Dennis Dieks, Wenceslao J. Gonzalez, Thomas Uebel, and Gregory Wheeler, eds. *New Challenges to Philosophy of Science.* Dordrecht: Springer Science+Media, 2013.

Blowers, Paul M.. "Gentiles of the Soul: Maximus the Confessor on the Substructure and Transformation of the Human Passions." *Journal of Early Christian Studies* 4 (1996), 57–85.

Brooke, John Hedley. *Science and Religion: Some Historical Perspectives*. Cambridge: Cambridge University Press, 2014.

Clayton, John. "The Otherness of Anselm." In *The Otherness of God*. Ed. Orrin F. Summerell. Charlottesville; London; University Press of Virginia, 1998: 14-34.

Clément, Olivier. *The Roots of Christian Mysticism*. Hyde Park, NY: New City Press, 1993.

Costache, Doru. "Experiencing the Divine Life: Levels of Participation in St. Gregory Palamas' *On the Divine and Deifying Participation*." *Phronema* 26 (2011), 9-25.

———. "Theology and the Natural Sciences in Palamas." In *God, Freedom and Nature*. Edited by Ronald S. Laura, Rachel A. Buchanan, Amy K. Chapman. Sydney, New York, Boston: Body and Soul Dynamics, 2012, 132–8.

Derrida, Jacques. *Acts of Religion*. Ed. by Gil Anidjar. New York and London: Routledge, 2002.

——— "How to Avoid Speaking: Denials," in *Derrida and Negative Theology*, ed. by Harold Coward and Toby Foshay. Albany: State University of New York, 1992, 73–142.

Eagleton, Terry. "Marxism without Marxism." In *Ghostly Demarcations: A Symposium on Jacques Derrida's Specters of Marx*." Ed. by Michael Sprinker. New York and London: Verso, 2008, 83–7.

Early Fathers from the Philokalia. Trans. E. Kadloubovsky and G.E.H. Palmer. London: Faber and Faber, 1954.

Eco, Umberto. *Foucault's Pendulum*. New York: Harcourt Brace Jovanovich, 1989.

Flory, Mark W. "The Art of Re-Integration: Sacred Art in the Light of Hesychastic & Other Perspectives." In *Orthodoxy and the Sacred Arts*, Sophia Studies in Orthodox Theology, Vol. 9, ed. By J.A. McGuckin. New York: Theotokos Press, 87–108.

Foltz, Bruce V.. *The Noetics of Nature: Environmental Philosophy and the Holy Beauty of the Visible*. New York: Fordham University Press, 2014.

Foucault, Michel. *The Essential Works of Michel Foucault, 1954–1984, Vol. 1: Ethics, Subjectivity, and Truth* (New York: The New Press, 1984.

———. *The Hermeneutics of the Subject: Lectures at the College De France, 1981–1982*. New York: Palgrave Macmillan, 2005.

Gregorios, Paulos. *The Human Presence: An Orthodox View of Nature*. Geneva, Switzerland: The World Council of Churches, 1978.

Gregory of Nyssa. *Commentary on the Song of Songs*. Trans. Casimir McCambley, OCSO. Brookline, MA: Hellenic College Press, 1987.

———. *Dogmatic Treatises and Other* Works. Ed. by Philip Schaff and Henry Wace. The Nicene and Post-Nicene Fathers, Vol. 5. Grand Rapids, Michigan: William B. Eerdman's Publishing Co., 1994.

———. *From Glory to Glory: Texts from Gregory of Nyssa's Mystical Writings*. Comp. Jean Daniélou. Trans. Herbert Musurillo, S.J. Crestwood, NY: St. Vladimir's Seminary Press, 1995.

———. *The Life of Moses*. Trans. by Abraham J. Malherbe and Everett Ferguson. Classics of Western Spirituality. New York: Paulist Press, 1978.

———. *On the Making of Man*. Trans. William Moore and Henry Austin Wilson. The Nicene and Post-Nicene Fathers, series 2, vol. 5. Grand Rapids, Michigan: William B. Eerdmans Publishing Company, 1994.

Haack, Susan. "Seven Signs of Scientism," *Logos & Episteme*, III (2012), 75–95.

Hadot, Pierre. *Philosophy as a Way of Life*. Ed. Arnold I. Davidson. Malden, MA: Blackwell Publishing, 1995.

Hadot, Pierre. *The Veil of Isis: An Essay on the History of the Idea of Nature*. Translated by Michael Chase. Cambridge, MA: The Belknap Press of Harvard University Press, 2006.

Handy, Michael. *No God, No Science? Theology, Cosmology, Biology*. Malden, MA: Blackwell Publishing, 2013.

Horujy, Sergey. *Practices of the Self and Spiritual Practices: Michel Foucault and the Eastern Christian Discourse*. Grand Rapids, MI: Wm. B. Eerdman's Publishing Co., 2015.

Jaeger, Werner. *Paideia: The Ideals of Greek Culture, Vol. II: In Search of the Divine Center*. Translated by Gilbert Highet. New York and Oxford: Oxford University Press, 1986.

Jones, Richard H. *Reductionism: Analysis and the Fulness of Reality*. Lewisburg, PA: Bucknell University Press, 2000.

Laird, Martin. "Apophasis and Logophasis in Gregory of Nyssa's *Commentarius in Canticum canticorum*." In *Studia Patristica* 37. Leuven: Peeters, 2001: 126–32.

———. "Gregory of Nyssa and the Mysticism of Darkness: A Reconsideration." *The Journal of Religion* 79 (1999): 592–616.

———. "The 'Open Country Whose Name is Prayer': Apophasis, Deconstruction, and Contemplative Practice." *Modern Theology* 21, 1 (January 2005): 141–55.

———. " 'Whereof We Speak': Gregory of Nyssa, Jean-Luc Marion, and the Current Apophatic Rage." *The Heythrop Journal* 42 (2001): 1–12.

Lee, Jung H. "Abraham in a different voice: rereading *Fear and Trembling* with care." *Religious Studies* 36 (2000), 377–400.

Lightman, Bernie. "Peter Harrison's The Territories of Science and Religion: A New Peter Principle." Science & Religion: Exploring the Spectrum. https://sciencereligionspectrum.org/long-reads/peter-harrisons-the-territories-of-science-and-religion-a-new-peter-principle/#more-819 (retrieved July 13, 2017).

Marcuse, Herbert. *One-Dimensional Man: Studies in the Ideology of Advanced Industrial Society*. Boston: Beacon Press, 1968.

Maximus the Confessor. "Four Hundred Texts on Love: First Century," in *The Philokalia*, Vol. 2, trans. by G.E.H. Palmer, Philip Sherrard, and Kallistos Ware. London: Faber Faber, 1981, §84, 62–63

Moore, Fr. Lazarus. *An Extraordinary Peace: Seraphim, Flame of Sarov* (Port Townsend, WA: Anaphora Press, 2009).

Nesteruk, Alexei V. *Light from the East: Theology, Science, and the Eastern Orthodox Tradition*. Minneapolis, MN: Fortress Press, 2003.

Nicolaidi, Efthymios. *Science and Eastern Orthodoxy: From the Greek Fathers to the Age of Globalization*. Baltimore: The John Hopkins University Press, 2011.

Ortner, Sherry B. "Theory in Anthropology since the Sixties." *Comparative Studies in Society and History* 26 (1984), 126–166.

Paden, William. *Religious Worlds: The Comparative Study of Religions*. Boston, MA: Beacon Press, 2005.

Palamas, Gregory. *The One Hundred and Fifty Chapters*. Trans. by Robert E. Sinkewicz. Toronto, Ontario, Canada: Pontifical Institute of Medieval Studies, 1988.

———. *Treatise on the Spiritual Life*. Trans. by Daniel M. Rogich. Minneapolis, MN: Light and Life Publishing Co., 1995.

———. *The Triads*. Trans. by Nicholas Gendle. Mahwah, NJ: Paulist Press, 1983.

———. "Topics of Natural and Theological Science and on the Moral and Ascetic Life: One Hundred and Fifty Texts, " in *The Philokalia*, Vol. 4 (London: Faber and Faber, 1995, 346–417.

Palmer, G.E.H., Philip Sherrard, and Kallistos Ware. *The Philokalia*, 4 vols. London: Faber and Faber, 1979–1985.

Peckler, Mark. "Imagination, Religious Practice, and World Transformations: Sophia, Heidegger, and Jacob Bohme's *The Way to Christ*." Ph.D. diss, Joint Ph.D. Program, University of Denver and Iliff School of Theology, Denver, 2009.

Placher, William C.. *The Domestication of Transcendence: How Modern Thinking about God Went Wrong*. Louisville, KY: Westminster John Knox Press, 1996.

Plato. *The Republic*. Translated by Tom Griffith. Edited by G.R. Ferrari. Cambridge: Cambridge University Press, 2008.

Podmore, Simon D. *Kierkegaard and the Self before God: Anatomy of the Abyss*. Bloomington and Indianapolis: Indiana University Press, 2011.

Ramfos, Stelios. *Like a Pelican in the Wilderness: Reflections on the Sayings of the Desert Fathers*. Trans. Norman Russell Brookline, MA: Holy Cross Orthodox Press, 2000.

Špidlík, Tomaš. *The Spirituality of the Christian East: A Systematic Handbook*. Trans. by Anthony P. Gythiel. Kalamazoo, Michigan: Cistercian Publications, Inc., 1986.

Ricoeur, Paul. *Freud & Philosophy: An Essay on Interpretation*. Translated by Denis Savage. New Haven and London: Yale University Press, 1970.

Sanguineti, Juan José. "Religious Faith, Natural Science, and Metaphysics."In M. Alai et al. (eds.). *Science Between Truth and Ethical Responsibility*. Switzerland: Springer International Publishing, 2015, 317–331.
Schatzki, Theodore R. "Introduction: Practice Theory." In The Practice Turn in Contemporary Theory, edited by Theodore R. Schatzki, Karin Knorr Cetina, and Eike Von Savigny. London: Routledge, 2001, 10–23.
Špidlík, Tomaš. *The Spirituality of the Christian East: A Systematic Handbook*. Trans. by Anthony P. Gythiel. Kalamazoo, Michigan: Cistercian Publications, Inc., 1986.
Tanabe Hajime. *Philosophy as Metanoetics*. Translated by Takeuchi Yoshinori. Berkeley: University of California Press, 1986.
Uždavinys, Algis. *Philosophy and Theurgy in Late Antiquity*. San Rafael, CA: Sophia Perennis Press, 2010.
Writings from the Philokalia on Prayer of the Heart. Trans. E. Kadloubovsky and G.E.H. Palmer. London: Faber and Faber Limited, 1951.

NOTES

1. See Mark W. Flory, "The Art of Re-Integration: Sacred Art in the Light of Hesychastic & Other Perspectives," in *Orthodoxy and the Sacred Arts*, Sophia Studies in Orthodox Theology, Vol. 9, ed. By J.A. McGuckin. (New York: Theotokos Press), 87-108.

2. A key argument of this essay is that science cannot achieve its intrinsic *telos* (goal) unless it goes beyond knowing about nature, and know nature as a transformative participant in it.

3. See Mark Peckler, "Imagination, Religious Practice, and World Transformations: Sophia, Heidegger, and Jacob Bohme's *The Way to Christ*" (Ph.D. diss, Joint Ph.D. Program, University of Denver and Iliff School of Theology, Denver, 2009). The "transcendental imagination" is "the ontological horizon of encounter with beings," and "as horizon of inner sense and the pure image of the schemata of concepts," it "does not create beings, but it does create the ontological horizon of encounter with beings." (p. 114)

4. This could explain the inherence of *logos* in *mythos* that Ricoeur notes.

5. For a good general survey of the "practice" turn in social scientific theory, see Theodore R. Schatzki, Karin Knorr Cetina, and Eike Von Savigny, eds., The Practice Turn in Contemporary Theory (London: Routledge, 2001), 10-23. For an influential reading of the contemporary history of social scientific theory, see Sherry B. Ortner, "Theory in Anthropology since the Sixties.," *Comparative Studies in Society and History* 26 (1984), 126-166. Still the best general introduction to practice theory in that of Joseph Rouse: Rouse, Joseph, "Practice Theory" (2007). *Division I Faculty Publications*. Paper 43. Of course, aspects of practice theory will be found in the analyses of any particular system of practice—so the potential sources include any theoretical reflection on any specific practice whatsoever.

6. See section IIB, below.

7. Theodore R. Schatzki, Karin Knorr Cetina, and Eike Von Savigny, eds., The Practice Turn in Contemporary Theory (London: Routledge, 2001), 11.

8. The degree of transformativity of a dominant paradigm therefore gives us an objective standard for measuring the health and, perhaps, viability of that paradigm. One of the key criticisms of Kuhn's theory of scientific progress (as crisis) is that it makes the standard for adjudicating "good" science utterly relative to dynamics of social and institutional power. Transformativity, as a standard, is not implicated in these dynamics, but provides an objective measure of them, according to their degree of transformativity itself. Thomas Kuhn is probably to thank for introducing the awareness of institutional and ideological impediments to scientific progress.

9. Michel Foucault uses this term to connote the mutual modifications between social and spiritual practices, on the one hand, and the formation of personal and communal identities, on the other. Subjectivation is the result of mental and physical activities (both private and social) that produce the subject as they produce knowledge. See Michel Foucault, *The Essential Works*

of Michel Foucault, 1954-1984, Vol. 1: Ethics, Subjectivity, and Truth (New York: The New Press, 1984), 195.

10. And all practices produce transformations of which the practitioner does not, and cannot sometimes, become aware. Whether these transformations are truly transcendent or otherwise is an important topic, but a different conversation.

11. For more on the characteristics of enlightened consciousness, see addendum #2.

12. Or at least, the co-creator of change. Jacques Derrida accuses the apophaticism of Dionysius the Areopagite (and other apophatic systems) with concealing an initiatory agenda. In his view, the secrecy of the apophatic strategy, and the directionality of the cataphasis-apophasis process, point to the positing of a transcendental signified, namely God as absolute being. This despite the uncompromising insistence of Dionysius (and others) that, being beyond Being, God cannot be conceptualized. I have argued elsewhere that the problem here is that the apophatic progression is itself conceived incorrectly. Its *telos* is not negation, nor some vaguely mystical union with God, but a reintegrative movement through and beyond enlightenment to transformativity itself. See Jacques Derrida, "How to Avoid Speaking: Denials," in *Derrida and Negative Theology,* ed. by Harold Coward and Toby Foshay (Albany: State University of New York, 1992), 73-142.

13. I will examine some of the important historical moments of this disruption later.

14. Bruce V. Foltz, *The Noetics of Nature: Environmental Philosophy and the Holy Beauty of the Visible* (New York: Fordham University Press, 2014), 3.

15. As we will see, the complexity thesis attempts to take into account the historical, social, institutional, theoretical, and other meanings of "science," as does Agazzi's systems theoretic approach. I will argue there that the complexity thesis tends to make the mistake of conflating a meta-analysis with its specific data. That is, it tends to rely on historical examples, which as historical are inductive and contingent. Instead of analyzing these examples from the perspective of a coherent and transcendental category, the complexity thesis tends to assume that the examples themselves prove what they merely indicate.

16. See Terry Eagleton, "Marxism without Marxism." In *Ghostly Demarcations: A Symposium on Jacques Derrida's Specters of Marx.*" Ed. by Michael Sprinker (New York and London: Verso, 2008), 83-7. I would argue that Derrida's attempts to preserve otherness fails for other, or at least, additional reasons. If otherness does not engage in reciprocal transformation with the self, it cannot even operate as a purely abstract ideal, as Derrida seems to want. This is the ultimate reason, as Eagleton charges, for Derrida's apparently weak political praxis: because otherness is so other as not to engage the same in transformation, both otherness and the same are voided of practical implication. To put it in terms of Derrida's analysis of the gift, if the gift is so purified conceptually that it cannot engage in economic relationships without losing its identity as gift, then the concept is denuded of practical import. Only by entering into economy can the gift be a gift; only through transformativity—the reciprocal transformation of gift and the gift economy—can it avoid reduction to human conceptuality (becoming what Marion would call an "idol").

17. Therefore, the teleological suspension of the ethical also brings into question Kierkegaard's well-known emphasis on freedom and individual choice. When the conditions for ethics are transcended, so are the conditions for making a rational choice—hence "leap of faith." The point here is that, like Plato, Kierkegaard's analysis seems to present an absolute difference between natural and enlightened knowledge. However, this diremption between the two types of knowledge is essential for transformation to take place.

18. See Jung H. Lee, "Abraham in a different voice: rereading *Fear and Trembling* with care," *Religious Studies* 36, 2000, 377-400 for a good consideration of several such interpretations. Lee argues for an "ethic of care," but I will argue that while this emphasizes relationship, it still does not take seriously impossibility. See also Simon D. Podmore, *Kierkegaard and the Self Before God: Anatomy of the Abyss*, Bloomington and Indianapolis: University of Indiana Press, 2011. Podmore argues that the otherness of God results from human sinfulness, a questionable thesis at best, ignoring as it does the traditional Christian distinctions of created/uncreated, infinite/finite.

19. Whereas the practices of enlightenment or illumination are spelled out in the Hesychastic texts, the practices of theosis are not. This is in part due to the fact that these teachings

would have been imparted in person, not by means of texts. It also points to the fact that these practices are essentially impossible to talk about, since they take human nature beyond its "normal" conditions. Practices like the contemplation of the Trinity name some of these practices, but obviously no definition will express the true import of the practice itself.

20. While the distinction between intellectual and practical (or heart-centered) forms of spirituality is fraught, it does help to explain one important aspect of spiritual progress, namely, that there are more rational and more intuitive or experiential aspects to it. For example, my first paragraph on Plato is incomplete and unfair, as it stands. For Plato, to achieve the transcendence of the canons of natural knowledge, i.e., enlightenment, one has also to transform their self. This is accomplished through the very rigorous Platonic *paideia*, including gymnastics, music, mathematics, logic, and dialectic. Through these processes, one transforms not only one's knowledge, but also one's being.

21. Richard H. Jones, *Reductionism: Analysis and the Fulness of Reality* (Lewisburg, PA: Bucknell University Press, 2000), 31-2.

22. Evandro Agazzi, *Scientific Objectivity and Its Contexts* (Cham: Springer International Publishing, 2014), 11.

23. Evandro Agazzi, "A Systems-Theoretic Approach to the Problem of the Responsibility of Science," *Zeitschrift für allgemeine Wissenschaftstheorie/Journal for General Philosophy of Science*, 18, (1987), 34.

24. Gregory's terminology is somewhat different from our own, since at his time science, philosophy, and theology were not yet distinguished in the way they are today. In addition, our distinction of "sacred" and "secular" simply did not exist for Gregory. When Gregory refers to "secular" learning, he is not referring to a kind of knowledge that had yet been separated from sacred knowledge. When Gregory speaks of philosophy, or Greek learning, or pagan learning, he is referring at once to what we would call philosophy *and* science *and* other forms of learning. As we will see, for Gregory, learning was a whole, and what distinguished "Greek" or "pagan" learning from "divine" learning was not its intrinsic nature, but its misuse or distortion of that nature.

25. Gregory Palamas, *The Triads*, Footnote 4, p. 118. Not to quibble, I should point out that the sentence could be interpreted more charitably as saying that Gregory denies the *necessity* of "secular studies" to the attainment of "deeper insight." However, even in this formulation, Gregory's actual position is obscured. Gregory *would* deny the necessity of secular studies, except that he is at pains to uphold their potential value. On the other hand, if by "deeper insight" is meant the knowledge that comes from illumination by the Holy Spirit, than certainly Gregory would agree that secular studies cannot be necessary, since they are utterly incapable of producing said knowledge. So, at best, the quoted sentence is extremely vague.

26. Gregory Palamas, *The Triads*, A.I.i.19, p. 27. According to the footnote in the *Classics of Western Spirituality* edition, "The "I" of the questions is meant to be the bewildered disciple who appeals to Palamas for guidance in face of the attack on the monks made by Barlaam and his supporters." (p. 117)

27. Palamas, *The Triads*, A.I.i., p. 25. The Christian monastic tradition arose within the socio-historical context of Greco-Roman culture and education. For this reason, the monastic authors utilized many literary and rhetorical devices learned from Greek pagan literature. In the case of Gregory's *Triads*, for example, we have at least the barest skeleton of a dialogue form. Dialogues can be a form of "indirect communication," to use Kierkegaard's term. Such communication can be informative, but conveying information is not its primary goal. Rather, indirect communication uses a variety of modalities (dialogue, dialectic, irony, questions, symbols, allegories, etc.) in order to put the burden of making meaning on the reader or listener (the audience).

28. Because by Gregory's time, such expressions of humility had become *topoi* of monastic literature, commentators today often ignore them. However, to do so is to ignore as well the performative (or indirect) nature of monastic treatises.

29. See Algis Uždavinys, *Philosophy and Theurgy in Late Antiquity* (San Rafael, CA: Sophia Perennis Press, 2010), 10. Uždavinys suggests that the dialogue form of philosophical writing attempts to emulate the "questions of the mystagogue and the responses of the initiate."

30. See Addendum 2, below.

31. To state it more precisely, Gregory worked within a tradition of monastic spirituality that had its own *topoi*, including traditional exegetical techniques (allegory, typology), rhetorical and hermeneutic tools, and literary forms. Gregory's contemporary audience, unlike his audience today, would not need to be informed of these *topoi*, since they were educated in and familiar with them.

32. Pierre Hadot, *The Veil of Isis: An Essay on the History of the Idea of Nature.* Trans. by Michael Chase (Cambridge, MA: The Belknap Press of Harvard University Press, 2006), 39.

33. Palamas, *The Triads*, §19, 25.

34. Palamas, *The Triads*, §20, 26.

35. Typical of all such practices, "discernment" can connote the exercise of discernment in regard to specific circumstances (such as demonic temptation, or giving spiritual direction), or it can connote the *telos* of such exercises, namely "perfect discernment," a discerning state of being. Of course, such a *telos* is not in any sense guaranteed by pursuing the practices, and most importantly, the more the practitioner attains discernment, the more that discernment transforms and is transformed by the practitioner. That is to say, there is a complex feedback loop between practitioner and practice.

36. See Paul Blowers, "Gentiles of the Soul: Maximus the Confessor on the Substructure and Transformation of the Human Passions," Journal of Early Christian Studies 4:1, (1996), 57–85. To be more precise, the passions, transformed through purification and enlightenment, will become completely oriented toward God.

37. Palamas, *The Triads*, §20, 26-7.

38. By "pride," Gregory does not mean the passion of any individual scientist, but the tendency of secular knowledge as a whole to serve human passions.

39. Palamas, *The Triads,* §16, 34. (my italics).

40. Costache, "Theology and the Natural Sciences in Palamas," in Ronald S. Laura, Rachel A. Buchanan, Amy K. Chapman (eds.), *God, Freedom and Nature* (Sydney—New York—Boston: Body and Soul Dynamics, 2012), 133.

41. Gregory Palamas, *The One Hundred and Fifty Chapters*. Trans. by Robert E. Sinkewicz (Toronto, Ontario, Canada: Pontifical Institute of Medieval Studies, 1988), §§1-2, 346-7.

42. Palamas, *The One Hundred and Fifty Chapters,* §§3-4, 347-8.

43. Palamas, *The One Hundred and Fifty Chapters,* §§9-14, 349-52.

44. See, e.g., Maximus the Confessor, "Four Hundred Texts on Love: First Century," in *The Philokalia*, Vol. 2, trans. by G.E.H. Palmer, Philip Sherrard, and Kallistos Ware (London: Faber Faber, 1981), §84, 62-63. We find this "sequence of sin" in the writings of many of the other Hesychastic masters as well.

45. Gregory Palamas, "Topics of Natural and Theological Science and on the Moral and Ascetic Life: One Hundred and Fifty Texts, " in *The Philokalia*, Vol. 4 (London: Faber and Faber, 1995), §19, 353-4.

46. Palamas, "Topics," §20, 354.

47. Palamas, "Topics," §28; 357-8. I will not address the vexed issue of the "actual" existence of demons here, but suffice it to say that Gregory would regard our modern skepticism about demons (and other "noetic" beings) as symptomatic of the very illness he is diagnosing!

48. Palamas, "Topics," §24, 356.

49. This paradox already existed for Plato: if the experience of the Good is so far beyond natural conditions of knowledge, how can it in turn perfect natural knowledge? In describing the conditions of natural knowledge, Gregory has already given hints as to the benefits of enlightenment in regard to natural knowledge.

50. Palamas, *The Triads,* A.I.iii.17, 34.

51. In the stage of illumination, the practitioner has achieved a state of transformation of natural knowledge through the opening of the spiritual senses. This results in the transcendence of the subject/object structure of natural knowledge in unitive knowledge, as well as the transcendence of the law of non-contradiction in what I have elsewhere described as "patristic dialectics." Apophasis, then, is not merely a linguistic/cognitive technique, but is the fruit of the practices of purification that result in these illuminative transformations. I examine the fruits of illumination (enlightenment) much more fully in my book on Hesychasm (forthcoming).

52. Palamas, *The Triads*, A.I.iii.19, 36.

53. In the monastic tradition, purification of the heart is the interior concomitant of the external means of purification, such as exile, solitude, and bodily ascesis (fasting, vigils, and sexual abstinence).

54. Palamas, *The Triads,* A.I.iii.18, 35.

55. Palamas, *The Triads*, A.I.iii.18, 35.

56. Gregory clearly states that negation by itself can liberate from created beings, but cannot "effect union with divine things." (Palamas, *The Triads*, A.I.iii.20, 33). Union requires purification of "the passionate part of the soul."

57. Gregory elaborates on the relationship between the intellect and the senses and the role of the imaginative faculty in Palamas, *The One Hundred and Fifty Chapters,* §§16-17, 353.

58. Palamas, *The Triads*, C.I.ii.2, 42.

59. Palamas, *The Triads*, C.I.ii.3, 43.

60. Palamas, *The Triads*, C.II.ii.9, 51. Much of the text of Gregory's *Triads* concerns the transformation of the senses and passions through bringing the spirit into the body, and the role of *ascesis* and suffering in this process of transformation.

61. Gregory Palamas, "Topics," §§146-7, 414-5.

62. Fr. Lazarus Moore, *An Extraordinary Peace: Seraphim, Flame of Sarov*, Port Townsend, WA: Anaphora Press, 2009, 11-138; see esp. 131.

63. Jacques Derrida, *Acts of Religion*, ed. by Gil Anidjar (New York and London: Routledge, 2002), 358-419.

64. Gregory also uses more defensible forms of reasoning, as I will discuss below.

65. Some of the following scientific criticisms of other, non-scientific, forms of reasoning derive from the brilliant insights of Charles Sanders Peirce.

66. Umberto Eco, *Foucault's Pendulum* (New York: Harcourt Brace Jovanovich, 1989), 56. It is arguable that there is more than a dose of the "lunatic" in many of these systems of thought. According to Eco, the "lunatic" "doesn't concern himself at all with logic; he works by short circuits. For him, everything proves everything else." (pp. 57-8)

67. While my concept of practice was inspired, in part, by Paden's concept of "world," the concept of practice goes much further than Paden's, and is deployed here in contexts which he may or may not recognize as valid.

68. William Paden, *Religious Worlds: The Comparative Study of Religions* (Boston, MA: Beacon Press, 2005), 7. For a fuller account, see Chapter 3: Worlds (51-65), and esp. 53-8.

69. Peckler, "Imagination," 109.

70. Peckler, "Imagination," 113.

71. Peckler, "Imagination," 114. While Peckler's description of imagination is Kant by way of Heidegger (made to serve an interpretation of Jacob Böhme's theosophy), the correlations between this description and the concept of khôra in Plato (or by way of Derrida's reading of Plato) are intriguing. Both are schemata of conditionless conditions, attempts to enunciate the ineffable other than constitutes (indeed creates; i.e., brings out of nothingness) selfhood.

72. Not to mention the resulting problems with the definition of "equal" in that statement.

73. Here, an important comparison might be made between transcendental imagination and the Hindu concept of purusha and prakriti.

74. Herbert Marcuse, *One-Dimensional Man: Studies in the Ideology of Advanced Industrial Society* (Boston: Beacon Press, 1968), 76-92. Marcuse describes how some forms of analytic philosophy promote the one-dimensional society, and contrasts it with dialectical, critical thinking.

75. On the other side, these recent epistemologies themselves assume—and to an extent, deny—a given cosmology, anthropology, and ethic. We might think that the practice of epistemology of this kind would dovetail with that of science, since, as I said above, the former in many ways acts as the ideological justification of the latter. Interestingly, the practice of epistemology, while it may model itself on the practice of science, does not entirely conform to the practice of science. This is not only because sometimes philosophers do not understand science quite as scientists do, but is also due to the fact that epistemology is its own practice, and therefore has its own *a priori* basis that, in turn, determines what it can regard as justification in the first place.

76. Sergei Horujy, *Practices of the Self and Spiritual Practices: Michel Foucault and the Eastern Christian Discourse* (Grand Rapids, MI: Wm. B. Eerdman's Publishing Co., 2015), 110. Horujy argues that the *telos* of a system determines the degree of transformation possible within it. He connects the "transcendence of the *telos*"—that is, the "ontological rupture" between human and divine—to the existence of practices of purification. A system whose *telos* is most absolutely other is one that emphasizes practices of purification.

Chapter Four

Envisioning Animality, Humanity, and Divinity with Inspiration from Georges Bataille

Donald L. Turner

1.

Although the mechanics of the system are often unrecognized or underappreciated by those who drive it, even brief reflection reveals that human activities on planet Earth have increasingly come to determine both *whether* non-human animals will exist and *how*, if at all, they will live. Many of these creatures exist simply to serve our interests, and they could easily perish should we change our minds or appropriate or ruin their habitats. Of course, how we treat non-human animals is related to how we think about them—as individuals and as members of particular species—and for the many millennia composing the long story of humanity's relationships with other animal species, we have thought and behaved ambivalently.

This ambivalence takes different forms through the ages. We see it today where, in the United States, for example, one winged species with a beautiful white head has been adopted as a national mascot and granted broad and powerful protection by federal law, while another with an equally lovely scarlet "comb" is torturously enslaved by the billions, its comb sometimes processed into human anti-wrinkle treatment. Members of a select few mammalian quadruped species have been welcomed into homes as members of human families, while those belonging to another species, by many accounts similar to humanity's "best friend" in terms of cognition, become a symbol of defilement and are, like their gallinaceous cousins, inserted into the factory farm machine.

All of this is bound up with basic aesthetical questions about human sensitivity to the beauty of other life forms—a capacity that is not required for one to recognize the inherent value of individuals of other species but that is helpful in this regard. My aim in this essay is to pursue some of these questions with respect to the phenomenon of human consumption of non-humans as food. The processes are as old as humanity itself and are responsible for many of the most atrocious abuses that non-humans suffer at human hands. Specifically, I draw upon the philosophy of Georges Bataille to chart ways to change our present thinking about three key concepts, humanity, animality, and divinity, in ways that might promote reform of current practices involving the non-human Others who die so that we may eat.

I find Bataille's work inspiring for numerous reasons. First, his work is notable for the seriousness and depth of his attention to questions about the meaning of animality and non-human animal phenomenology. From the opening sentence of his first full-length book under his own name in which he cites "the secret lives of animals" as the first in a list of "the really important things in life," to the last philosophical treatise published during his lifetime where he provides his final summary of his thoughts about how humanity gradually emerged from its pre-Paleolithic origins to distinguish itself among Earthly animals in terms of its awareness of its own mortality, its eroticism and eventual subjection of sexuality to rules, and its capacity to express its ultimate concerns and ideals through art, his work sheds light on both what we and other animals have in common and what distinguishes us. Moreover, his description of the general trajectory of human thought about animality from prehistory to today is accurate and illustrative; it is easy to see how the dynamic he describes plays out in the realm of human carnivorous activity, which is as clearly involved in the prehistoric images with which Bataille's story of humanity's emergence begins as are the themes of sexuality and death that occupy center stage in his philosophical writings.

Finally, although Bataille does not overtly stake out this domain, his writings point toward at least two ways humans might, for both their benefit and that of their non-human kin, reform their thought and practice. The first is to reinvigorate aspects of the early human mindset that recognizes non-human animal forms embody the sacred and that treats their lives with the appropriate reverence and the process of killing and eating them with the appropriate solemnity. The second is for humans to recognize new aspects of divinity within themselves, reframing vegetarianism and veganism, for example, as exercises in the types of expenditure by which, according to Bataille, humans attain glory, majesty, and sovereignty.

2.

For Bataille, it all begins with art, and a full understanding of the ways humans have thought about the sacred, other animals, and themselves must extend back into prehistory to consider some of the oldest artistic representations known to exist—the paintings on the walls of caves, made by humans living tens of thousands of years ago in places such as Lascaux in France and Altamira in Spain, where the three themes of humanity, animality, and divinity are all at play. Although scholars debate their exact significance, most of the images are bound up with the hunt, mainly depicting the animals who were the painters' prey. In numerous essays and sections of longer works, Bataille expresses wonder at the effort required to reach some of these long-hidden places and the care inherent in the production of the images of early humans' prey animals with such beauty and realism that, to my admittedly untrained eye, puts the cave artists in league with competent artists of today[1] [Figures 1-2].

Basing his supposition on his reading of the cave paintings, Bataille infers that these early humans must have had such deep love and respect for their prey animals that they divinized them; to them, "animal nature was sacred".[2] He supports this thesis by pointing to various theriomorphic characters (such as those from the Egyptian pantheon [Figure 4.3]) common to ancient religions and to statements by contemporary peoples whose modes of life are closer than ours to those of prehistoric people, such as the Navajo, whose tenet is cited by Bataille: "Wild game is like man, only more godlike".[3] Bataille speculates that for the prehistoric hunters who were positively dis-

Figure 4.1. Photograph of Lascaux animal painting.

Figure 4.2. Reproduction of a bison of the Cave of Altamira, Cantabria, Spain.

posed toward their prey and toward animality in general, to kill, transform into foodstuff, and eat these divine embodiments would have been scandalous, even criminal—a sense he sees reflected in rituals of thanksgiving and apology that accompany contemporary tribal societies' carnivorous activities.[4]

Interestingly, the cave painters rarely depict human beings, and when they do, they do so quite unrealistically. For example, the human figure in the Lascaux painting is drawn with a simple outline, the face concealed behind a non-human animal mask [Figure 4]. From this, Bataille infers in the artist a feeling of suspicion of or discomfort with his/her own species. As Bataille sees it, our prehistoric ancestors revered the animal condition from which humanity had only recently and tentatively emerged, and he speculates that the animal masking, also seen in images such as the horned personage sometimes referred to as "The Sorcerer" [Figure 5], might have been meant to elevate and honor the human by incorporating elements of non-human animality into its form. The development of the human species from that point forward features a gradual withdrawal of divinity from the animal world and a denial of the animal within the human being. I do not see this as an irreversible process, however—a point I hope to make clear in figures 4.4 and 4.5.

In any event, we have at least a partial picture of prehistoric consciousness, a depiction of the relationship of the hunter with his prey, painted for us by Bataille. But what about the consciousness of the non-humans so reverently depicted on the cave walls? What were their lives really like? Given

Figure 4.3. From left: Anubis, Set, Horus, and Hathor.

this art precedes the development of agriculture, at least insofar as non-human animals' interactions with humans are concerned, it seems reasonable to assume the creatures who were these peoples' prey lived their own lives, each individual pursuing its own *telos* until the day it fell in the hunt. To be sure, death, bloody and painful, was delivered then as now, and, admittedly, in some instances, today's proximate causal injury might be less painful and bring death more swiftly than in ages past. But for prehistoric prey animals, death came suddenly and constituted only a tiny fraction of the lifespan. The scandalous conversion of a living animal, embodying the sacred and possessing a subjectivity and agenda of its own, into a thing or a tool occurred relatively quickly at the end of an otherwise independent life.[5]

3.

Like our Paleolithic ancestors, we consume animal flesh, audaciously converting the animal that originally exists for itself into a tool or foodstuff. The similarity between the contemporary mentality and that of our predecessors ends here, however. In practice, while it's reasonable to assume that for Paleolithic hunting tribes, participation in the predation was widespread (among the males at least), only a small percentage of meat-eaters in today's United States play a role in the killing, with most of us entering the process several steps later, after the army of low-wage workers our purchase will help pay has performed the grueling, dirty work. Like our predecessors, we seem to be the kinds of beings for whom violence against other animals feels

Figure 4.4. Shaft Scene with "Bird Man" and bison from the Lascaux Cave, Dordogne, France.

transgressive, but unlike the people who struggled through this, leaving traces of the drama on those cave walls, we move the process away from view and take other measures to avoid the struggle that we would face if we were we stronger. Ultimately, by distancing ourselves from the violent actions, we avoid a troubling sense of scandal and shame and thus the need to perform rites of apologies and thankfulness to honor those we consume.

Along the way, of course, contemporary humans have largely lost their capacity to recognize anything sacred in the lives of other animals; divinity has relocated. As Bataille explains, for our Paleolithic ancestors, the animal was divine, and humanity, newly emerging and unsure of itself, looked for ways by art and rite to bridge the novel gap between themselves and their sacred animal heritage. We, on the other hand, have inherited a way of thinking in which, if any creature manifests the divine, it is ourselves, a view reflected in the first chapter of the Judeo-Christian creation myth, which declares that humans alone were created "in the image and likeness" of God, with other creatures divinely ordained to live under human dominion from the start and, shortly thereafter, being officially recognized as legitimate foodstuff given by God to humans as plants had previously been.[6] In our humanity-animal-divinity triad, humanity has replaced animality as the element associated with divinity, and animality, instead of humanity, is the objectified and excluded third. In other words, the sensibility driving the cave paintings has been inverted.

We can see traces of these changes in the images accompanying consumer meat products. Whereas the cave painters took pains to render the real

Figure 4.5. "The Sorcerer" from Cave of the Trois-Frères, Ariège, France. Wellcome Library, London

animal present in their images, modern practice obscures food animals' real context, for this is what we most do not want to see, lest our conscience tremble. For example, we change the names: cow becomes beef; pig becomes pork. In addition, the final purchased product bears little resemblance to the living form behind it, and no one dares put a snapshot or realistic image of the actual miserable, imprisoned animal on the package or in the advertising lest consumers refuse to buy. Instead, we more often see a smiling cartoon caricature representing a system that, from a utilitarian perspective, causes vastly more evil than the prehistoric way of obtaining meat [Figure 6].

Our prehistoric ancestors could portray their food animals accurately and realistically because their predatory activity reflected the human virtues of humility, sympathy, and gratitude, while the disjunction between the simplistic images of modern industry's food animals and the real animals who languish under that system betrays the bad faith of the humans involved in predation today. If the prehistoric hunter viewed his predation as a scandal, with the death of a divinity requiring an apology, what would he think of the modern factory farm? But modern people do not think as prehistoric people did. Instead of theriomorphic gods, we have a hierarchical chain of being in which chickens and swine embody links so distant from divinity that they do not matter at all in their own right; they are morally relevant only for the ways our treatment of them impacts members of our own species.

Figure 4.6.

These changes have brought huge differences in the life quality of food animals. As I noted previously, prehistoric prey animals lived in the wild, pursuing their own ends, free of human control until the day of their death. In contrast, contemporary factory bred and reared food animals are subject to human ends for most or all of their lives, as the scandalous objectification process swallows the entirety of their existence. It is here, rather than in the fact of killing itself, that I locate the main wrongfulness of dominant meat production practices. For the benefit of all animals, human and non-human alike, I would like to help reverse these trends, and in the sections that follow, I draw further on Bataille's philosophy to envision two dietary modes as possible alternatives to the modern industrial meat production system.

4.

The first of these dietary modes involves a return to a more Paleolithic model, elements of which are practiced today among small numbers of people, although in ways that might be strengthened and better coordinated among those who practice them and more effectively promoted among those who do not. One of these elements is the recognition and ceremonial acknowledgement of the wonder, beauty, and sacredness of the animals we eat, calling for new prayers and rites—or old ones revived—to honor them. Here the care taken by our prehistoric cave painting ancestors serves as a model. We would, of course, develop our own rites and rituals, ones that speak more clearly to contemporary society. The rites would represent and inspire the kind of respect that would clearly rule out factory farming, in which the animal is subject to a level of degradation unthinkable for something divine.

The second mode is the closer participation of meat eaters in the lives and deaths of the animals eaten. The reality of the modern factory farm is so well hidden that those who consume its products can do so with the illusion of clean hands and a clear conscience. At some level, I think, most people realize present practices are wrong, and if they had to personally inflict upon the victim all the suffering these practices involve, the suffering would be reduced dramatically.[7] Increased knowledge about the lives and deaths of our food animals is likely to correlate with our increased regard for them. This is well-known in hunting societies; successful hunters must know their prey, and Bataille remarks anecdotally that hatred of their prey is foreign to the minds of hunters with whom he is familiar.[8] In contrast, people who pay for factory farmed meat typically do not dwell on the reality of the food animals' lives, and the system includes elements designed to keep that knowledge from the minds of eaters, including misleading advertising, as mentioned above, and measures known as "ag-gag" laws such as the one recently passed in Iowa, which establishes criminal penalties for the production, possession, and distribution of photographic or audio records of events that transpire at commercial animal facilities.[9] In my own more utilitarian approach to questions of animal death, considerations about reducing the total balance of suffering are combined with considerations of the percentage of the victim's life that is spent pursuing its own ends versus the percentage in which the predator forces it to submit to her own needs and wishes. This leads me to posit that in an ideal world, individual meat eaters with the capacity to do so would hunt and kill their own wild game.

If we consider a standard U.S. diet with meat at nearly every meal, the math on the number of animals and number of acres of woodland or pastureland required for them does not work out for free ranging species such as deer or elk, but the available land mass would sustain the present standard meat consumption if the animals were "farmed". I recommend a middle path here, suggesting carnivores cut their meat consumption to, at most, a third (preferably to something like a seventh or a tenth) of the "meat at nearly every meal" level, eating a meat dish only two to three times a week. Maintaining woodlands for *some* hunting would ensure habitat for a range of species to live and pursue their natural ends until the time at which they fall prey in the hunt, while limiting meat consumption would allow the environment in which meat animals live and die to look more like "the wild", better embodying the ethics of respect for nature, with aesthetic value for humans coming as a bonus. For individuals in this world, the hunt, accompanied by appropriate rituals, would be honorable but rare, as an individual eating meat once or twice a week could provide his yearly meat consumption with one deer, and an elk could feed a family. (Different species, of course, would have to be substituted in different locales, but the logic remains the same.) Meat-eaters would approach their predation more like our ancestral hunters who, although they could not suppress their predatory instinct any more than they could suppress the sexual activity necessary for their species's continued life, could at least subject it to rules and give the act of killing a "sacramental character" with "the significance of a religious violation."[10]

This is the ideal situation for carnivorousness, I think, but I have no illusions that such a model will be quickly or widely embraced. Even so, the general idea of honoring our meat animals with more reverence and ritual and bringing the consumer back into closer connection with the lives and deaths of food animals would improve things for those species we raise. If meat eaters had to look at the lives of their food animals frequently, if they had to visit the farm, if they personally had to select and slaughter a pig or a cow, they might harbor more respect for the individual animal, translating into better lives for human and non-human animals alike. Despite logistic hurdles, this seems preferable to the present system, in which a family might eat a whole cow's weight in beef over the course of the year, but served up in tiny anonymous chunks from hundreds or thousands of different animals, with maximum separation between consumer and consumed and no sense of the misery and degradation the process involves.

5.

Beyond valorizing certain elements of prehistoric meat provision, Bataille's philosophy offers resources for grounding a theologically oriented vegetarianism that might form the general context within which such carnivorousness occurs. Let me be clear: this is not a return to prehistoric ways but a new way of thinking, focused not on allowing other animals to manifest divinity to us, but on our manifesting divinity to them.

To illustrate how this works, I turn to Bataille's economic analysis in the three-volume opus, *The Accursed Share*. Here he argues that in addition to what he calls a "restricted economy" or an "economy of scarcity", in which the focus is on the production of wealth and where a shortage of resources necessitates judicious allocation of assets for specific, carefully chosen ends, a complete assessment of human economy must include attention to the "general economy," a concept which posits that while individual organisms frequently face resource shortages (and thus reasonably adopt an economics of scarcity), "[o]n the surface of the globe, *for living matter in general*, energy is always in excess".[11] In this passage, he is pointing to the consumption of resources and the processes by which societies and individuals expend or even squander a surplus wealth that cannot be used for productive growth. This wealth is, in essence, the concretization of energy, of which there is normally a "surplus [that] must be dissipated through deficit operations," violating the norms of the more rational and utilitarian "restricted" economy by "sending it [the wealth] up in smoke".[12] Fueled by the sun, which continuously expends without return more energy than living matter can absorb, the "general economy" will always contain a surplus that will be lost to the system or spent for non-productive purposes.

Bataille sees examples of wasteful consumption throughout the many circuits of energy on Earth. Human sexuality is one such circuit, with immense stores of energy and time devoted to non-productive ends. Ironically, given my goal in this section, one of his main examples involves the earthly food chain. Bataille writes that "the sexual act is in time what the tiger is in space"—namely, a site for the expenditure / consumption of excess energy.[13] In this metaphor, the sun throws out vast riches of energy; plants absorb all they can, but this is, of course, only a fraction of the whole. Herbivores absorb what they can from the vegetation, again only a fraction of what the sun produces. Then the tiger, in "an immense squandering of energy," consumes the herbivore and all the solar units its life represents. By accomplishing what the laws of the general economy require, the carnivore is valuable. However, in this section I want to suggest something different—namely, that Bataille's analyses of sacrifice and gift-giving might help to frame vegetarianism / veganism as the construction of an "admirable monument" to the power of our love and concern for our fellow Earthlings, the non-human animals who would become food. I propose that we reflectively adopt the expenditures Bataille metaphorically represents as sacrifice and gift giving as part of our dietary thought and practice.

To explain what he means by the expenditure of sacrifice, Bataille turns to the Aztecs, who, according to Bataille, were "poles apart from us morally" in that they were as concerned with consumption and sacrificing as modern societies are about production and working.[14] They were almost constantly at war, but not for some rational, productive reason such as the expansion of their realm or the acquisition of slaves. Instead, the goal, religiously motivated, was to obtain victims for sacrifice. Rather than make productive use of the wealth slaves represented, by destining them for ritual slaughter, the Aztecs withdrew them from the cycle of productive work—the level of profane utilitarianism—and elevated them to a sacred place befitting an offering to the gods.

Then, to illustrate the expenditure represented by gift-giving, Bataille looks to the practice of potlatch among Northwest Native Americans; in the most exemplary form of potlatch practiced by chiefs and sovereigns, the objects given to others are luxury items (as opposed to sacrifice, in which items are more mundanely useful things such as food animals or slaves). He notes that in cultures practicing potlatch, a person's place in the social hierarchy is reflected in and determined by her capacity to expend resources with no eye to possible profit—the higher the value of the luxury goods that person sheds, the higher his prestige and rank. At a fundamental level, this represents not the power to take but the power to give; by dispensing wealth without return, the giver obtains glory. At these sacred moments, the giver resembles that symbol of profitless expenditure *par excellence*, the sun, transcending the profane, utilitarian world, affirming her sovereignty and approaching divine status.

The ambiguity of non-human animals' ontological status complicates the decision about whether to frame vegetarianism as sacrifice or potlatch—or both. With meat's status as foodstuff in view, it looks very much like sacrifice, but when we consider most people's nutritional needs can be met without meat, animal flesh as food looks like a luxury. Reflecting this ambiguity, vegetarianism can include elements of both mechanisms, which have in common an ability to facilitate profitless expenditure according to the principles of the general economy.

On the one hand, in sacrifice, as Bataille sees it, useful items are withdrawn from circuits of profane exchange and the utilitarian world of things in which everything serves ends foreign to the thing itself and are returned to the original condition of animal immanence. I suggest viewing vegetarianism as sacrifice writ large. Whereas ritual animal sacrifice involves removing one particular animal from the food production circuit, adopting vegetarianism might be seen as performing a similar gesture for the whole class of non-human animals, withdrawing them all from this circuit in one fell swoop. By giving up the power to make productive use of the other animals that she might enslave and consuming to fuel her bodily maintenance, the vegetarian performs a potentially religiously significant sacrifice by forgoing meat.

On the other hand, adopting vegetarianism also resembles the kind of extravagant giving that conveys prestige. Again, in societies practicing potlatch, rank is determined by the ability to give, and the chief is the one who is rich enough to give the most—to have the last word in the rivalry of giving by being rich enough to give so much that the recipient cannot return the gesture in an act of "one-upmanship". I suggest viewing vegetarianism along these lines as a means to enhance our sense of personal splendor. When aromas of well-cooked meat provoke carnivorous cravings, we might choose to feast instead on the pride of conquering these forces, forgoing nutritional or gustatory benefit while generating admiration for ourselves as too physically powerful to need to consume animal flesh to survive and too spiritually noble to participate in the industrial desecration of animal life. Moreover, insofar as there are no recipients who can return the gesture in kind, adopting vegetarianism is like performing the ideal potlatch, which, as Bataille recognizes, "could not be repaid".[15]

6.

In terms of how we should view the relationships between the elements of the triad of concepts covered in this essay, humanity, animality, and divinity, I think that the best approach is to see the choice between a Paleolithic model, with animality and divinity on one side of the divide and humanity on the other, and a modern model, with humanity and God on one side and animality on the other, as a false dilemma. Instead, the way of thinking I am trying to entertain sees divinity pervading both the human and non-human animal realms. On the one hand, viewing ourselves as divine, we might leave carnivorous activity behind altogether as something we see ourselves as too powerful to require. On the other hand, viewing the animal as sacred would require consigning the factory farm to the proverbial ash bin of history and returning to a more Paleolithic model in which we have the strength to face directly any fellow creatures whom we are to kill and consume, bringing about their deaths more intentionally and reverentially.

Similarly, perhaps the choice is not between Paleolithic hunting, which modern people do not have time to pursue day after day anyway as our ancestors did, and some kind of pure vegetarianism or veganism, which, if pushed to its absurd conclusion, would require voluntary starvation as the only way to completely avoid living at the expense of others' deaths. I say this because, as minimal reflection reveals, almost no one eats without causing other creatures to die. Tractors that plow soybean fields kill rodents, and the vegetarians who purchase tofu made from those soybeans annihilate insects with their scooter windshields on the way to the store. Failing to recognize this might make it seem too easy for a vegetarian or vegan to think his hands are cleaner than they really are, and someone who thinks as Bataille does might see embracing vegetarianism without recognition of these realities as just another way for a modern person to deny and thus betray shame about her animal nature – the same kind of shame that is evident in a horror of death or a denigration of sexuality. Recognizing that our lives are sustained by processes that cause others' deaths helps move a discussion about food ethics beyond a simple argument about whether or not eating other animals is morally acceptable in general to a more nuanced conversation about the particular conditions of *all* life forms whose existence is affected—and in many cases produced—by our dietary practices.

Along these lines, I envision a dietary mode in which we neither turn a blind eye to the reality of animal death, in the manner of most factory meat consumers, nor pretend to live without requiring any violence to other living beings, as do some myopic vegans. Perhaps we might blend the two dietary modes described above—that of the glorious vegetarian and the prehistoric noble hunter—to embrace a mostly vegetarian diet, reducing or preventing as much non-human animal suffering as possible, while punctuating this general abstinence with periodic violations of the rule, participating as closely as possible in the death of the animal we are to consume during the violation. Carnivorousness of this type could be framed within an overall context of peace and renunciation of violence toward non-human animals that rules out the abhorrent, whole-life abuses of factory farming. Instead, the hunting experience would take on the air of transgression, lifting the general prohibition of violence for a temporary return to and honoring of the animal condition that our prohibitions, according to Bataille, enable us to transcend.

BIBLIOGRAPHY

Bataille, Georges. *The Accursed Share*, vol. 1. Translated by Robert Hurley. Brooklyn, NY: Zone Books, 1991.

———. *Cradle of Humanity: Prehistoric Art and Culture*. Translated by Michelle Kendall and Stuart Kendall. Brooklyn, NY: Zone Books, 2005.

———. *Erotism: Death and Sensuality*. Translated by Mary Dalwood. San Francisco, CA: City Lights Books, 1986.

———. *Theory of Religion*. Translated by Robert Hurley. Brooklyn, NY: Zone Books, 1992.

NOTES

1. Many of Bataille's writings these images are collected in Georges Bataille, *The Cradle of Humanity: Prehistoric Art and Culture* (Brooklyn, NY: Zone Books, 2005) although he also discusses them in Georges Bataille, *Erotism: Death and Sensuality* (San Francisco, CA: City Lights Books, 1986), 74-75, 83-85.
2. Bataille, *Erotism*, 84. For similar comments, see Bataille, *Cradle of Humanity*, 55.
3. See Eveline Lot-Falck, *Les Rites de chasse chez les peuples sibériens* (Paris: Gallimard, 1953), quoted in Bataille, *Cradle of Humanity*, 131.
4. See, e.g., Bataille, *Eroticism*, 81-85 and *Cradle of Humanity*, 127-132.
5. For one of his clearest analyses of this process, see Georges Bataille, *Theory of Religion* (Brooklyn, NY: Zone Books, 1992), 39-42.
6. I do not mean to suggest that this is the only or final word on this subject for the Judeo-Christian tradition, which also includes a range of saints who experienced God through intimate friendship with non-human animals and many voices, both historically and today, representing various forms of religious vegetarianism. But the myth is revealing, and it helps undergird a mindset according to which factory farming is acceptable.
7. Admittedly, the situation is complicated insofar as people who cut chickens' throats for a living might be better at it than the first-time slaughterer dispatching birds in the yard. However, I think those regularly killing their own food animals would seek education and try in the long run to cause as little suffering as possible, while the machine of agribusiness, oblivious to animal suffering, routinely sees animals awake and aware for extremely gruesome procedures.
8. Bataille, *Cradle of Humanity*, 51.
9. Iowa Code Ann. § 717A (West 2012).
10. Bataille, *Erotism,* 74.
11. George Bataille, *The Accursed Share,* vol. 1 (Brooklyn, NY: Zone Books, 1991), 23.
12. Ibid., 22.
13. Ibid., 12.
14. Ibid., 45-61.
15. Ibid., 70.

Chapter Five

"Seeing Nature Whole in Fragments" or "Whole in Nature Seeing Fragments"

Sharon Adams

> One is necessary, one is a piece of fate, one belongs to the whole, one is in the whole—there exists nothing which could judge, measure, compare, condemn our being, for that would be to judge, measure, compare, condemn the whole...*But nothing exists apart from the whole*! —Friedrich Nietzsche[1]

> Human life—indeed all life—*is* poetry. It's we who live it, unconsciously, day by day, like scenes in a play, yet in its inviolable wholeness it lives *us*. It composes *us*. This is something far different from the old cliché 'Turn your life into a work of art'; we are works of art—but we are not the artist. —Lou Andreas-Salomé [2]

> I am part of the part that was once everything. —Johann Wolfgang von Goethe[3]

INTRODUCTION

Frank Seeburger begins his essay "Seeing Nature Whole in Fragments" with a reference to two key phrases from a lecture by Martin Heidegger. These two phrases become pivotal for Seeburger's argument as he develops it, or rather more accurately allows it to unfold, moving backward and forward through time and location, referring to various thinkers as they are situated within their particular philosophical, social or political timeframe and context. His line of reasoning ends in our own place in time, one that some claim qualifies as a new epoch, the Anthropocene, where human activity is impacting the earth's climate and ecosystems, forever. Even as we are all a part of the whole of humanity, who has happened to have polluted our environment, causing mass extinctions, and effecting climate change, we must ask ourselves how, and why?

Through his line of reasoning that begins with a juxtaposition between two ways of thinking, Seeburger opens a way for an answer to these questions, allowing his readers to decide for themselves which of the two phrases and their philosophical as well as practical implications, makes the more sense heading into the future. He gives his readers an invitation to join him, further, in developing an aesthetics of nature that can inspire, if one is open to it, not only to face nature in general, or as a whole, but specifically, as one human face faces a nature which faces them back. In his essay, though a type of mutual movement, philosophical, aesthetic, and theological approaches to nature and the divine interchange. Through the interchange comes understanding as to the how and why of the predicament we are in. Only then, through the ability to see and be seen, can one thereby *do* something about the wholesale destruction of nature.

In Seeburger's unique *making* of an aesthetics of nature, he presents interlocking themes linked with, among others, the interaction between the individual and nature (subject and object), notions concerning fragment in relation to poetry, healthy agon between humans and nature in/through love, and in reference to art and nature, a loving struggle between world and earth. At the conclusion of his essay he describes the enlivening practice of praying with icons from the Greek Orthodox Christian tradition, describing the process in which one is able, in prayer, to go beyond the icon as object, that is as idol, in order truly to see a divine as and in nature. His reference to a going beyond withstands the temptation to stay with (literally focus on) the object, succumbing, as appears to be the tendency in human nature, to the ever present longing for stability, stasis, security or *home,* thereby fixating on the art as refuge or escape. If one is able to set their sights beyond the idol as Seeburger suggests, there comes not only an ability truly to see icon/nature as divine, but also to be open to be seen by nature. There arises in mutual recognition a loving embrace.

Seeburger's aesthetics of nature holds promise in breaking through traditional philosophical approaches to an environmental philosophy and theology, certainly as they stem from reductionist and materialist presuppositions toward nature. Rather than making a contribution to refute a prior argument, philosophical or otherwise, he leads his readers toward an opening that may enable real change from past patterns of destruction, even, or perhaps especially, ones within the academy and religious traditions as they function in our world today. I find his perspectives most persuasive, as he engages in a Heideggerian pedagogy, starting as and ever remaining a beginner, that is with beginner's mind, and inviting his readers to join him in his reasoning.

This methodological approach leads his reader to a realization not so much through vilifying the/an enemy, in thinking or otherwise, arising as it usually does from a polemical approach out of revenge or as Nietzsche's notion of *ressentiment*. It instead arises out of willing, active, and engaged love. This is how there can be change, not only because humans may love nature, in the end, but also because they finally are able to feel her love. How then, arises the obvious question, could we as humans during the Anthropocene, continue to kill her?

There are many rich and promising avenues to explore in responding to Seeburger's essay in ways that attempt to answer this question. In order to begin developing a particularly feminist approach to the philosophy of religion following the lines of Seeburger's aesthetics of nature, I assume traditional perspectives on philosophy, theology, art and nature should be adjusted according to the new epoch in which we live. The new epoch as Anthropocene carries a double burden of weight, one for realizing the damage we've done, and two, as a call to duty to address the problems we've created. It wouldn't do for a true re-evaluation and transformation just to be opposing a masculinist mindset with a feminist one, or to turn upside a master-slave dialectic where "women rule," but by playing by the same rules as the masculinist game. Nor would it be beneficial to turn to a mythical past in order to re-envision nature as the "eternal feminine" in an utopian Goddess-epoch, seeking a matriarchal pre-history when women ruled from which to model our current lives on.[4] I would like rather to highlight a feminist philosophy of religion that transcends the polemic of gender binaries not by ignoring them, but by turning them upside down, but also around.

I start out by pursuing some references to Jewish theologian and philosopher Franz Rosenzweig, namely from his *Star of Redemption*, and in particular his references to Friedrich Nietzsche. I view Rosenzweig, as with Heidegger, as diverting from the philosophical thinking that tends to value a systematic totalizing of the whole at the expense of the singular individual being. Tracing Rosenzweig's thought in reference to Seeburger's aesthetics of nature leads, at least to some extent, toward Rosenzweig's reference to Nietzsche as the first philosopher as man. In his *Star of Redemption*, Rosenzweig identifies Nietzsche as the first able fully to break from the western philosophical tradition, as fragment of whole in Seeburger's terms, leading toward what Rosenzweig would consider a philosophy as life, rather than philosophical thinking as suicide.

Even as Nietzsche holds promise, according to Rosenzweig, to function as first philosopher as man, breaking from the history of western philosophy culminating in German Idealism, Rosenzweig also describes Nietzsche, in the end, as not being able to fulfill the full measure in this promise. Rosenzweig's final reference to Nietzsche in the *Star* conflates Nietzsche with his creation, Zarathustra in fact, in his/their "decline and fall."[5] It is at this very point of supposed failure, however, of Rosenzweig fusing author/artist in mutual creation, and describing their fall in mutual demise, that I find Nietzsche *a la Zarathustra* may be resurrected.

In order to substantiate this perspective I utilize French philosophers Sarah Kofman's and Eric Blondel's approaches to Nietzsche's style of writing in relation to poetry and metaphor to provide a lens from which to develop a feminist philosophy of nature and religion according to an unique and timely methodology; Kofman holds to a historical trajectory similar to one Rosenzweig, Nietzsche, Heidegger and Seeburger address, reaching back to the Presocratics (following Heraclitus) and culminating, as seems to be a common consensus, with Nietzsche. Her approach in particular may add some insight not seen before within the context of this particular mix of thinkers, and may offer a means to help redeem Nietzsche/Zarathustra's thought, from what Rosenzweig would identify as a "fallen state."

I conclude my exploration by turning to a play written by Lou Andreas-Salomé only recently translated into English: "The Devil and his Grandmother." Interestingly enough, Matthew Del Nevo and Gary Winship, the authors of the introduction to Andreas-Salomé's *Sex and Religion: Two Texts of Early Feminist Psychoanalysis* take note of the fact that "The Devil" was written at the same time as Rosenzweig's *Star*, and that both texts follow the primary and fundamental movement, from death to life.[6] "The Devil," as possible "reflection" of the *Star*, however, functions not as any simple dialectical model of antithesis to traditional theology/philosophical approaches to nature as thesis, or in terms of any type of synthesis of two opposing points of view. Instead, through her playful seeming narrative Andreas-Salomé adds complexity and nuance to a feminization of nature that makes her readers think. She brings in a psychic dimension in her retelling of a mythical narrative drama, depicting an approach to devotion and love not completely unlike what Rosenzweig is able to do in his *Star*. I seek to explore ways in which Andreas-Salomé's play does more than mimic a face of God within traditional mainstream theological and philosophical notions of the divine, including aspects of Rosenzweig's Star of David as the face of God. We must have theological grounds to help support what I would call a Nietzschean aesthetics of nature; this is what a particular reading of Andreas-Salomé's "The Devil" can give us, to fill in the gap, by having the gap speak.

There may be something useful within Andreas-Salomé work that can add to Seeburger's exploration in helping to develop a philosophical approach to an aesthetics of nature that stands apart from philosophical materialism prevalent in the past, as well as in contemporary philosophical approaches to the environment, ethics related to the environment, and a theology that supports both. I suggest as much, focusing on Andreas-Salomé's "The Devil and his Grandmother" in toying with some of her "elemental" symbolism and characterization, as well as writing style (that of a poem as play and film) in addressing issues revolving around the fragment as whole, a healthy struggle between earth and world, and a theological perspective on how her poem as art can help to highlight a feminist voice speaking not for all women, as a whole (which as Seeburger demonstrates in his thought, is impossible), but as one singular voice able to provide helpful critique on gendered notions of earth in relation to God/good and the Devil/evil.

PART I: AN AESTHETICS OF NATURE: HOW DO WE BEGIN?

Seeburger begins his essay "Seeing Nature Whole in Fragments" with a reference to Martin Heidegger's 1929 lecture "What is Metaphysics," where Heidegger makes a distinction between "the whole of beings" (*das Ganze des Seienden*) and "being in the whole" (*das Seiende im Ganzen*). Seeburger describes Heidegger's phrase "whole of beings" as meaning "everything that in any sense 'is'…all put together to form one single totality."[7] This phrase Seeburger associates with a view of the world as a totalized entirety. It implies a way of seeing and being in the world that eliminates any space as opening needed for individual agency or liberation.

Seeburger juxtaposes this phrase with another phrase from Heidegger's speech, "being in the whole," which Seeburger interprets as "given to us whenever any single being at all is given to us." This second phrase, "being in the whole," in contrast to the first emphasizing a whole as totality, elicits a type of wholeness found, as it turns out, only in and through separate parts/fragments. Seeburger refers to these two ways of viewing the whole (as two ways of being in the world), emphasizing the second as an individuality come alive only as fragment of the whole. This is a view that leads to hope, ultimately, for a more stable and sane future, with humans taking the the helm in doing something to change their behavior and identify viable ways of living in order to help save the earth altogether.

Seeburger leads his readers from fundamental insights in Heidegger's references to being in relation to the whole, toward a model for realization that Seeburger in fact includes in a quote from Meister Eckhart at the very beginning of his essay, even *before* the beginning, in fact, in his epigraph: "The eye through which I see God is the same eye through which God sees me." Utilizing a critical process of a double seeing and double reading, we find in Eckhart's quote a representation of a moment of transformation inherent in a realization that there is only ever one individual specular awareness (both human and divine eyes are the same; there is only one eye). Such a realization arises in relationship between individual and the whole, and comes about only through one's capacity of being a fragment (through a process of "fragmenting" as opposed to "totalizing"); only then can one come to the recognition that there was never really any separation from the whole in the first place. One realizes there is then, in Eckhart's as well as in Nietzsche's words from *Twilight of the Idols, Or How To Philosophize with a Hammer* I include in my own epigraph, no-thing a-part-from the whole.

Through references to multiple writers, ideas, and events over time, Seeburger explicates how a preliminary understanding of the two ways of looking at the world relates to various ways of thinking and living. He weaves a thread of reasoning building toward articulating an aesthetics of nature that may encourage his readers to "do" something in relation to an underlying and overwhelming (if faced directly) predicament we are in, that is the threat of extinction. In the end, his perspectives give hope in continuing to find a will to live fully in what appears to be the death throes of nature.

As Seeburger asserts in his essay, such a realization cannot be "taught" in the sense of an overly didactic pedagogy that Seeburger and Heidegger appear to steer clear of; nor can it be "made" in the sense of praxis or poesis. Following through with some of Seeburger's insights, it comes only after doing philosophy as a beginner, that is just by "doing" it rather than thinking or talking about it, and then wait for a response. Seeburger alludes to what the response may be at the end of his essay, with his description of the Greek Orthodox Christian practice of praying with icons. Within this practice, art plays a role in seeing nature, and allowing nature to respond.

Seeburger describes "seeing" icons in prayer as a going beyond the object toward a clearing necessary to experience love in reciprocity. This seeing nature double provides the groundwork for an aesthetics of nature that plays off of a "doable" and "double seeing," two themes Seeburger utilizes as at least two threads through a Labyrinth in which, to hearken to the Yeats' poem *Second Coming*, "the center cannot hold." The center as such, may be nothing other than a realization of reality Seeburger refers to in a type of "seeing beyond" an art object when praying with icons. Seeburger lays out a particular outward journey toward home, I suggest, in order not so much to "teach" his audience about such a return home, but again, in how he describes in Heidegger's approach to the study of philosophy as a beginner, by actually doing it and inviting his readers to follow.

Given our current global environmental crisis I suggest that gazing directly at nature is a place we can't bear to look. Our earth is dying, and it is nearly impossible to face. Even as there could be hope in seeing beyond nature, toward a type of divinity in and through Nature, certainly there arises need for caution not to mistake a *seeing beyond* with an escaping of the fact that nature is losing ground, literally. Such an obvious death I will be exploring in terms of a dying (Mother) earth, speaks to the very origin of western philosophical inquiry culminating in German Idealism according to Jewish theologian and philosopher Franz Rosenzweig. For Rosenzweig, western philosophical thought going back to the Presocratics is founded upon seeking to avoid facing death.

PART II: ROSENZWEIG ON NIETZSCHE AND BEYOND

As Seeburger begins his essay with the distinction Heidegger makes between "the whole of beings" and "being in the whole," Rosenzweig also, in relation to the history of western philosophy, refers to two approaches to life and the world. Rosenzweig identifies what he calls the "cognition of the All," beginning with Plato/Socrates and leading to Hegel's Idealism, with an attempt to know the "whole of beings" as a totality. If we follow Rosenzweig's trajectory alongside Seeburger's line of reasoning, then at least according to Rosenzweig in the beginning of Part One of the *Star*, Friedrich Nietzsche stands against the all-pervasive move in the history of western philosophy toward knowing the all in a totalizing sense.

In Rosenzweig's first reference to Nietzsche in the *Star*, Nietzsche breaks the mold; he cuts himself free from history, as a fragment of the whole:

> Till now, all philosophical interest had revolved around the knowable All; even man had been entitled to be an object of philosophy only in this relationship to the All. Now, facing this knowable world, there rose another independent reality: the living human being; before the All there rose the One who mocked all totality and all universality, there rose the 'Unique One and his property.' It was not in the book of that title—which was after all only a book—but through the tragedy of Nietzsche's life that this innovation was then encrusted into the riverbed of the evolution of the conscious mind.[8]

It is not through philosophical writings, but in Nietzsche as a human, that he stands a-part for Rosenzweig. We may find the Nietzsche that Rosenzweig presented to his readers is akin to what Seeburger heralds as a precursor of one who overcomes the "polemical—from the Greek *polemos*, 'strife' or 'war'" in favor of a more Nietzschean poetic *agon*-er, from the Greek *agon*, that is as a struggle between friends, in/through love. This is a type of love as fraternity and sorority with philosophers as human beings playing with strife, competition and struggle within an agreed upon context of mutual understanding.[9] Rosenzweig may likewise address the notion of being in the whole as fragment, and art as struggle, not so much reminiscent of *polemos* that rests on any either/or, friend or enemy binary, but instead in viewing art's role through another lens, as Heidegger's use of the Greek word *aletheia*, "that is, in the revealing of beings."[10] It is clear that such an opening cannot be forced; it has to open on its own.

Art coming about in agon-y can more aptly describe Nietzsche's type of struggle with his "higher self"/soul, within his philosophy; it is struggle akin to Jacob wresting with an angel, in the sense of Nietzsche as philosopher as *man*, struggling, *not* with an enemy, but with an angel/or representation of the divine (his soul) in the hopes of coming to the place/space that Jacob did, seeing God face-to-face. Here American poet and essayist Adrienne Rich helps decipher such a struggle in commenting on the somewhat ambiguous as well as ubiquitous Biblical passage of Jacob wrestling with an angel. Rich comments on a struggle inherent in poetry that helps distinguish the difference between the terms *polemos* and *agon*:

> In the Biblical story of Jacob's struggle, the person he contents with remains mysterious. He claims to be God, but the tradition prefers him as a messenger. Either way, it's not just another man. If we follow the story to this extent, the struggle is between human and metahuman, poetry and experience, poetry and material forces. Our angel, then, is not the 'the angel of history,' but the messenger from the lived reality of the barricades facing the poetry, the individual carrier of poetry. Within this messenger poetry has to wrestle every day.[11]

Struggle can lead toward a break/fragmentation of the whole, not necessarily toward destruction in/as demise but in order to create an opening/pathway truly to see. According to the Biblical narrative, after the storm of struggle is over and Jacob's hip is permanently damaged, Jacob is able to see God face-to-face (the eye which sees God is the eye which sees him); his name becomes Israel, one who struggles with God, and the place he called "Peniel, saying, "it is because I saw God face to face, and yet my life was spared" (NIV, Genesis 32:30).

Rosenzweig addresses a similar juxtaposition of struggle between poets, saints and their souls, indicating Nietzsche is one able to struggle toward overcoming, uniting poet, philosopher and saint: "Here there arrived a man who knew his life and his soul like a poet and obeyed their voice like a saint, and yet he was a philosopher." Rosenzweig elaborates upon Nietzsche's pivotal role in history, as the first philosopher as poet: "But he himself, who, in the metamorphoses of his figures of thought, himself metamorphosed . . . it is he whom none who must philosophize can henceforth bypass." Rosenzweig concludes: "For Nietzsche, there was not this separation between the height and the plain in his own Self, he went entirely his way, soul and mind, man and thinker remaining one until the end."[12]

In commenting upon the role of art in society, Seeburger also praises Nietzsche: "Art—real art…Nietzsche puts it in his 'Prologue' to his *Thus Spoke Zarathustra*, from Zarathustra's great speech about the 'Overman' and the 'Last Man'—is there to call us to do our duty."[13] This call to duty cannot come out of compulsion, fear or resentment, but only out of love, out of *making* love in terms Seeburger suggests. Likewise, may we consider Seeburger's comment upon Jean-Claude Milner's use of the notion of "fragmentizing as opposed to totalizing," in being able, as Seeburger quotes Milner, "to incite rebelliousness," to evoke "the very resistance that can break us out of the totalizing strategy that otherwise entraps us."[14]

Seeburger's is a call to duty for action, but through art, "real art, and not just some hurdy-gurdy sort of entertainment to give every one 'their little poison for the day, and their little poison for the night'…" [15] The reference in Nietzsche's quote here hearkens to Goethe's *Faust* and the ambiguous nature of the poison Mephistopheles offers in tempting Faust. Just enough poison numbs and inebriates, too much can kill. Rosenzweig also cautions against drinking the poison, in relation to the seduction of the cognition of the all:

> It is presumably necessary for man to disengage once in his life. Like Faust, he must for once bring the precious vial down with reverence; he must for once have felt himself in his fearful poverty, loneliness, and dissociation from all the world, have stood a whole night face to face with the Nought. [16]

There is ambiguity concerning the reference to "drinking the brown juice"; in some ways it can reflect Nietzsche's poison as opiate to avoid the fact that we all will die, of facing and embracing transience and death, but again there is an implication that too much of the poison will do more than distract, it can actually kill.

It is crucial to note that the supposed encounter of death we read about in Goethe's *Faust* is not so much an encounter with death itself, as in loss of life, but instead it is an encounter with a type of despair that comes with anticipating death, in considering how in death all one's plans and ambitions will have become thwarted. [17] This is the real fear which creates anxiety; it is one that is perhaps more difficult to face than the fact that we all will die. The anxiety comes forward in facing not so much death itself, but the notion of death, the Nought. Even as it is risky to do so, the rewards of opening to the Nought can enhance the possibility of overcoming past habitual patterns of behavoir that inhibit and even harm our planet and ourselves.

This fear or anxiety in the face of death, according to Rosenzweig, philosophy addresses "by weaving the blue mist of its idea of the All about the earthly. For indeed, an All would not die and nothing would die in the All. Only the singular can die and everything mortal is solitary. Philosophy has to rid the world of what is singular…" [18] Appropriately, Rosenzweig associates death, rather the fear of death, with the earth: "Without ceasing, the womb of the indefatigable earth gives birth to what is new, each bound to die, each awaiting the day of its journey into darkness with fear and trembling. But philosophy denies these fears of the earth." [19] We can read the phrase "fears of the earth" in terms of human fear of death, and birth, which will ultimately lead to death in nature, but we can also read the phrase as the *earth* having fears—which go beyond our own projections on birth, life and death. There is tremendous fear of facing death; again the fear or denial stems not just from dying, which could be seen in and throughout nature's various cycles, whether they last a few days or years, but in experiencing the particularly human experience of despair that accompanies life in anticipation of death.

True/pure art can function to provide bridges between living life in denial of life (as a form of suicide), and facing death in/as anxiety in the face of the Nought and thereby overcoming the all-to-human resistance to accepting transience. Such a role resonates with contemporary English author and art critic/theorist Jeanette Winterson's ideas, particularly in reference to the role of "true artist" in a society she aptly describes as "drugged on reproduction."

Winterson's objection to an objectification of art, of idols of all sorts, supports Seeburger's progression toward developing aesthetics of nature that *sees both ways*. Winterson writes:

> The true artist is connected. The true artist studies the past, not as a copyist or a pasticheur will study the past, those people are interested only in the final project, the art object, signed sealed and delivered to a public *drugged on reproduction.*

Winterson's reference to a true artist rings in harmony with Seeburger's reference to Paul Celan's pure poetry, as well as with Nietzsche's notion of agon/struggle and Rosenzweig's perception of Nietzsche as man and philosopher:

> The true artist is interested in the art object as an art process, the thing in being, the being of the thing, the struggle, the excitement, the energy, that have found expression in a particular way. The true artist is after the problem. The false artist wants it solved (by somebody else).[20]

The notion of true art, pure art, in a world system such as ours, becomes clearer through an understanding of Winterson's play on words. She describes the art process as a chiasmic relationship between the two phrases "thing in being," and the "being of the thing"; the relationship includes struggle, but also excitement and energy that cannot be reproduced for mass consumption, but only "in a particular way." Winterson may give us a key to deciphering, on a deeper more complex level, Heidegger's similar chiasmic structure between the "whole of beings" and "being in the whole"; through the chiasmic idea we may now see a clearing that opens to us, as a revealing or disclosure.

Winterson also uses the term "art objects," such that the phrase can be read two ways. Highlighting its ambiguity, it can be read not unlike Seeburger's references to seeing nature and an aesthetics of nature. One may read the phrase art objects, with the word *objects* functioning as a plural noun, or one may read the phrase art objects, and read the word *objects* functioning as a verb, as in an art that *objects*. Winterson acknowledges a similar movement in her critique of the type of thinking aligned with Heidegger's phrase "the whole of being":

> We know that the universe is infinite, expanding and strangely complete, that is it lacks nothing we need, but in spite of that knowledge, the tragic paradigm of human life is lack, loss, finality, a primitive doomsaying that has not been repealed by technology or medical science. The arts stand in the way of this doomsaying. Art objects. The nouns become an active force not a collector's item. Art objects.[21]

The way art *can* object Winterson explores through a pedagogical exercise similar in some respects to a *true* praying with icons. She describes it as follows: "Supposing we made a pact with a [original] painting and agreed to sit down and look at it, on our own, with no distractions, for one hour."[22] Winterson suggests something may open up through this exercise, even while sitting within the museum setting with its obligatory "tearoom/toilet/gift shop," where one "experience [es] paintings as moving pictures, out of context, disconnected, jostled, over-literary, with their endless accompanying explanations, over-crowded, one against the other, room on room." In her exercise, Winterson suggests that if one can withstand the increasing discomfort, distraction, invention, and even irritation, one may find that "when the thick curtain of protection is taken away; protection of prejudice, protection of authority, protection of trivia, even the most familiar of paintings can begin to works its power." What can happen, according to Winterson, is that the viewer can "fall in love."[23]

Winterson's approach to art may not identify the seeing and being seen of art explicitly in terms of a divine; her approach does play off of a neo-Marxist critique of capitalist society, however, we find in Walter Benjamin's essay: "The Work of Art in the Age of Mechanical Reproduction." As Benjamin describes the lack of aura from the original artists within the replication of art within modernity, Winterson is able to see that original art loses its aura in the reproduction of images. A particular specular approach to art as icon possible in sitting with an original painting can be unbearable for some, yet, as Winterson asserts, it is only through sitting with the art in silence that one can not only see, but actually in the spirit of Seeburger's reference to the Greek Orthodoxy practice of praying with icons, be seen. Perhaps this experience can open the observer to what is really there, but it takes opening first to what is beyond the static image, and this happens only through a taste of the Nought.

Again, according to Rosenzweig, Nietzsche appeared able to face the Nought as a man who emboldened his philosophy with life, who broke from the past to overcome the totality involved in Heidegger's phrase "whole of beings." Rosenzweig describes such a model of overcoming: "Man became a power over philosophy—not man in general over philosophy in general, but one man, one very specific man over his own philosophy." To reiterate, man, or a man, this one particular man, has broken free of history, as a fragment of/as the whole, "Man, in the simple oneness of his own being, in his being which was established on his last name and his first name, strode out of the world that knew itself as a thinkable world, strode out of the All of philosophy."[24]

Nietzsche appears in Rosenzweig's assessment as philosopher and poet in terms of the type of pure poetry Seeburger associates with Paul Celan, that is as poetry able to function as art that, in the helpful phrasing of writer Jeannette Winterson, *objects*. To expand upon such a poetry as *art that objects*, we find in Rosenzweig's assessment of Nietzsche not so much a response or reaction to political and social injustice through a slave revolt in morality, that is as a form of *ressentiment*,[25] but instead in real active engagement and inspired action. In contrast, a slave morality fits nicely with the political rhetoric of someone such as Carl Schmitt, as Seeburger points out, setting up a poverty mentality of us vs. them, friend vs. enemy.

To help move us forward in finding relevance for our own attempts to use art that objects as pure poetry, especially in our current attempts to develop a philosophical and theological aesthetics able to aid in bringing awareness and inspiration for change, Nietzsche's critique of action arising from ressentiment comes to the fore. To help clarify a distinction, I turn to Adrienne Rich's reference to the difference between protest poetry that attempts to address the "whole of beings" acting as Nietzsche would assert, out of revenge, but then to be freed up to act as a truly creative agent in alignment with "being/s in the whole" in what Rich associates with dissident poetry.

Rich quotes a passage from American poet James Scully in his observation concerning the difference between protest poetry and dissident poetry:

> Most protest poetry is conceptually shallow. I think of the typical protest anthology: poems in opposition to the Vietnam War or the coup in Chile, ecologically concerned or antinuke poetry (with a few devastating exceptions, mainly Japanese), even poems sympathetic to workers (notably those that focus on workers' oppression...ignoring the exploitation that necessitates the oppression). Such poetry is issue-bound, spectatorial—rarely the function of an engaged artistic life...It tends to be reactive, victim-oriented, incapacitated... it seldom speaks the active rage or resolution of...oppressed or exploited people...Dissident poetry, however, does not respect boundaries between private and public, self and other. In breaking the boundaries it breaks silences: speaking for, or at between *with*, the silenced; opening poetry up, putting it in the middle of life...It is a poetry that talks back, that would act as part of the world, not simply as a mirror of it."[26]

Dissident poetry, as fragment ("act [ing] as part of the world") breaks through the totality of political discourse. It functions as art that does more than just mimic; it holds the promise to transform its readers/viewers.

At least in terms of Rosenzweig's initial reference to Nietzsche in the beginning of Part I of the *Star*, Rosenzweig views Nietzsche as a dissident, having not succumbed to the Faustian wager: "Philosophy had promised to give him compensation in the form of mind in return for selling it his soul, and he no longer took this compensation seriously."[27] The reference to drink the "little poison for the day, and their little poison for the night," perhaps implies he sells his soul, at least as in Thomas Mann's Faustus account, only to lose his senses after his epoch of power, fame and success ran out.[28]

Yet, Rosenzweig complements Nietzsche as one who was liberated from the cognition of the All. Nietzsche was able to function along the lines of an integrated model for philosopher with a soul, a pure and dissident poet; for, even as Nietzsche's particular ideas may have faded, his mode of philosophy had not: "It almost doesn't matter today what he philosophized about. The 'Dionysian' and the 'Übermensch,' the blonde beast, the eternal return – where have they gone?"[29] These "teachings" may have been forgotten, but what remains is the "true" philosophical courage and urge to continue on one may hope to see in future beginning and budding philosophers.

Ultimately for Rosenzweig, however, it was not Nietzsche, but Goethe who came closest to the prototype for Rosenzweig's "New Thinking"; albeit not quite at the level of the Jewish and Christian openness to revelation, and therefore true love, that is love in the spirit of Seeburger's reference to making love as it arises in an indirect and organic fashion.[30]

PART III: ROSENZWEIG ON NIETZSCHE: SAVIOR OR DEMON?

The first reference Rosenzweig makes about Nietzsche in the opening pages of Part I of the *Star* has Rosenzweig both praising and condemning Nietzsche: "But none of those who now feel the urge to philosophize can any longer by-pass the man himself, who transformed himself in the transformation of his images."[31] Even as Rosenzweig describes Nietzsche as a guidepost for any true philosopher coming after him, Rosenzweig also appears to fuse Nietzsche's philosophy with Nietzsche's mental image of Zarathustra in terms Rosenzweig will ultimately critique. It is as if the fusion itself, of Nietzsche mixing his thinking with his creation, discounts Nietzsche's philosophy as having value, at least in terms of Rosenzweig's New Thinking. Nietzsche may be one who set the preconditions for Rosenzweig's hopeful future, but his use of a poetic style disqualifies him in being one able to usher in a New Thinking and being.

In the opening pages of Part III of the *Star*, Rosenzweig explicitly explores ways that Nietzsche was not able to become more than a tightrope or bridge between the old philosophy/philosopher and the new. Actually crossing over the bridge Rosenzweig called the Johannine completion, and it is Goethe who personifies such a "post" modern Christianity of the future:

> Goethe's life is really a walk on a ridge between two abysses: in every moment of his life he succeeded keeping within his depths on firmly established enduring ground. Any other whom the arms of divine love did not catch, and let him take flight into the eternal would necessarily have plunged into one of the two abysses that gape below both edges up to which everyone must henceforth climb to give vitality to his life.[32]

Rosenzweig highlights Goethe's prayer as especially fitting for the future "post" modern Christianity. Even as Goethe was pagan, and his prayer was really to his own "fate," Rosenzweig praises the prayer in reaching towards ultimate becoming. He describes this prayer in ways that resonate with ideas associated with "being in the whole":

> In these prayers, the parts which, in their conjunction, make the 'Part'' impartible, each have become alive in their own right.[33] Out of body and soul there is thus created a Whole: man has become an 'individuality,' and in his prayer for what he already possesses, his own fate, this wholly individual aspect is not also vitalized as such; it incorporates itself in All and yet does not cease to be individual.[34]

Rosenzweig views Goethe as one able to pray to his "fate" as ultimate potential, this is a prayer that goes beyond the mere paganism of Nietzsche. It goes beyond Nietzsche because it transcends the notion of individual fate.

Rosenzweig grapples with the notion of the prayer of the individual, in how this singular "'Here I stand' of the individual God awakened soul," as a type of "being in the whole," can have an impact on the world:

Already because these are multiple orders, they cannot without further ado be the same as the one divine order. For this, they would have first have to be the same among themselves. And they are not so, as long as each of the many still goes back to a lonely prayer of a lonely soul.[35]

Rosenzweig asks the question: "Would there be a power in prayer that could tyrannically intervene in the course of the world sprung from God since Creation?"[36] Rosenzweig insists that the manner in which the individual prayer can maintain efficacy in the face of the whole/world comes about only in praying for the farthest most universal, *as well as*, having love for the nighest, nearest, closest being/entity. This is a prayer that coincides with the coming forth of the Kingdom of God, and its timeliness is essential: "Only the prayer which is offered at the proper time will not delay the advent of the kingdom of heaven."[37] Even as Rosenzweig describes Goethe's prayer as "the prayer of the nonbeliever," which "cannot itself protect itself from...stumbling into the dual falsifications of time, in the too-late of the sinner and the too-soon of the fanatic," he also notes that Goethe "grasps the precise moment of the right time, the favorable time, the time of grace." Rosenzweig concludes: "Only now does the kingdom of God really have its advent within time."[38]

In contrast to Goethe's prayer to fate Rosenzweig praises, comes a reference to Nietzsche as Zarathustra appearing to have succumbed to both of the two abysses Goethe avoids, the sinner's prayer which is always too late, and the fanatic's prayer which is always too early:

A little illustrated notice is set up on the ridge: following the example of Zarathustra's decline and disappearance, it shows how one can become an immoralist who breaks all the tablets and a tyrant who does violence to his neighbor as to himself for the sake of the second nearest, or to his friend for the sake of the new friends – sinner and zealot in one person.[39]

Rosenzweig conflates Zarathustra and Nietzsche, and dismisses them as both a sinner whose prayer comes too late, and a fanatic whose prayer comes too early. Rosenzweig disparages Nietzsche as a sinner "smashing old tablets," who, as Rosenzweig notes in his section entitled "The Sinner's Prayer," fails to show gratitude for his very existence. Instead of recognizing that all life is a gift from God, the sinner makes requests as if there was "something as yet unfulfilled." According to Rosenzweig, the sinner should have made his request "before his creation;" for, "once created, he can only thank for his Own.[40]

A sinner's prayer does not come at the right time, and in gratitude. As the prayer of the fanatic, Nietzsche's prayer also does not *come at the right time*. In contrast to the sinner's prayer, which Rosenzweig describes as "remaining in the sphere of the Own,...exclude[ing] itself from the wealth of love which is awaited and needed by the moment of the desirable time," the prayer of the fanatic, "longs to accelerate the advent of the kingdom, to have it come before its time".[41] Rosenzweig writes of the fanatic's prayer as

> seek[ing] to capture the kingdom forcibly at the point, which the searchlight of his prayer shows him as the next one but which never is closer than the next-but-one. His prayer and his love wither for him and so in the end he himself has also withdrawn from the moment, full of grace. . . .[42]

For Rosenzweig, Nietzsche, in seeking the next-but one, misses the very individual directly in front of his face: "For the sake of the distant One which the moment of the prayer indicates as essential to him, everything other than this one, everything nighest, must be forbidden to him."[43] He, like the fanatic "long[s] for the next-but-one and thereby exclude[s] themselves from the host of those who advance along a broad front, covering the face of the earth bit by bit, each of them conquering, occupying, inspiring his nighest."[44] For Rosenzweig, Nietzsche suffers from projecting a "whole of being" attitude, missing the urgency of the call, as well as the call of intimacy of the one closest to him.

Goethe was still a pagan according to Rosenzweig, if even a pagan on the road to redemption; he was not therefore able to receive of the love that allows him passage over and through the threshold of love in being able to see the Divine face-to-face. Whereas both Nietzsche and Goethe fall short in terms of Rosenzweig's New Thinking, due to their "paganism," Goethe in the beginning of the third and last part of the *Star* is the "father" of the future Johannine Christianity as "pure son of this earth." Again, even as Goethe fulfilled the role of "man" as microcosm, who "never misses the mark, is never too late nor too early...," who prays "ever in the desirable time, the 'time of grace,' and is fulfilled as fast as it is prayed," in the end Nietzsche misses the mark through his "merging with a character of his creation" in eventual demise.

To reiterate, Rosenzweig critiqued Nietzsche/Zarathustra in "becom[ing] a sinner and a fanatic in one person," a sinner who prays too late, presumably, because he did not pray the prayer of gratitude, and also Rosenzweig found Nietzsche to be as the fanatic who prays too early because, unifying Nietzsche and Zarathustra as one, he/they jumped ahead to the next-than-one forcing the coming of the kingdom before its time. It is in this very reference, however, with Nietzsche using a poetic representation of Zarathustra as philosopher, poet, and saint, that I argue Nietzsche may in fact have presented his philosophy in no better way.

PART IV: NIETZSCHE/ZARATHUSTRA AS DEMON OR POET, PHILOSOPHER AND SAINT?

Rosenzweig appeared not able to appreciate certain of Nietzsche's poetic writings and use of metaphor. Rosenzweig's critique of Nietzsche appears typical of what the French philosopher Eric Blondel notes in his essay "Nietzsche as Metaphor" concerning philosophical analysis of Nietzsche throughout history:

> Most critics have insisted on considering Nietzsche's 'poetic' and metaphorical style of writing as either the simply and often tasteless ornamentation of philosophical prose produced by a good-natured poet, or as the kind of decoration this is favored by 'men of letters,' but that philosophers try desperately to forget.[45]

Many philosophers and theorists in the mid to late 20th century address Nietzsche's style in more favorable terms.[46] French philosopher Sarah Kofman, for one, I would assume may find Nietzsche able to fulfill his promise of integrated philosopher, poet, and saint. She reinforces the notion that it is in fact through Nietzsche's literary style of writing he is able to do so.

Kofman notes that Nietzsche makes use of metaphors in his philosophy, "even if it risks being confused with poetry":

> This is a confusion, however, that Nietzsche does not find regrettable: The opposition between philosophy and poetry reinstates metaphysical thought; it rests on the fictional separation between the real and the imaginary, and on the no less fictional separation of different 'faculties.' To speak in metaphor, then, is to have language regain its most natural expression, its 'most accurate, most simple, most direct' style.[47]

Kofman does not regret Nietzsche conflating philosophy and poetic metaphor. Her support of metaphor in philosophy highlights its ability to be evocative and persuasive, have personality, and address issues that would otherwise lose their vitality. Through her interpretive lens, Nietzsche as Zarathustra is not some flight of fancy, but a true and accurate representation of a fiction truer than what science, mathematics and technology could depict.

Eric Blondel likewise addresses Nietzsche's style in terms that redeem "Nietzsche/Zarathustra." For Blondel, as for Kofman, Nietzsche's use of poetic images and metaphor did not stem from some narcissistic or fantastic diversion, but instead it was intentional: "Because of his deliberate use of polysemantic metaphors rather than neutral concepts, it would seem more judicious, or perhaps even more philosophic, to ask if Nietzsche's 'style' does not necessarily embody a philosophical choice."[48]

Pulling from one of Nietzsche's own philosophical *fragments*, "The Last Philosopher," Blondel quotes Nietzsche: "Art treats appearance as appearance, therefore it does not seek to deceive; it is true." Furthermore, Blondel comments on Nietzsche's thought:

> Knowledge has no privileged position: on the contrary, 'there is no intrinsic knowledge without metaphor'... Nietzsche does not understand by this that everything is tantamount to the illusions of fantasy: rather, the real manipulation of appearances—lying appearances at that—has been performed by science, morality, and religion—precisely what had formally passed themselves off as 'the truth.'[49]

Nietzsche, identifying the "lie" concerning traditional and conventional notions of truth rallies against a type of totalizing discourse, Seeburger describes as "one that strives to reduce all to one—a self-important, bombastic, and totalizing discourse that over-fills itself with discursive content, striving to leave no room for anything else to voice itself or be heard." [50] Kofman quotes Nietzsche on the aesthetic dimension of philosophy in similar terms:

> The fact that an *indemonstrable* philosophy still has value—most often even more so than a scientific proposition—stems from the aesthetic value of such a philosophizing, that is, from its beauty and sublimity. Philosophizing still remains a work of art, even if it cannot be demonstrated as a philosophical construct. [51]

Kofman, like Nietzsche, Rosenzweig and Heidegger before her, refers to philosophical epochs in history. Her reference to Nietzsche in terms of the nature of epochs is "typological" rather than evolutionary: "Nietzsche finds a veritable break between the dawn of philosophy and its later development: the Presocratics belong to a rare type; they are irreducible to any other. [52]

Kofman also makes a distinction between two styles of philosophy, juxtaposing Aristotle with Heraclitus and Socrates with Dionysus, as follows:

> 'The pathos of distance' that separates two kinds of life that have always been with us: one is flourishing, superabundant, and projects its own excesses onto things and embellishes them with it. The other is degenerative, and it can only impoverish the world by reducing it to the narrow and ugly bounds of the concept—it does this out of spite against itself and out of resentment toward life. [53]

Kofman following Nietzsche wants to "portray or depict the Presocratics, then, but not by caricaturing them, not by 'reducing' them in stature."[54] Kofman refers to a passage in Nietzsche's *The Last Philosopher*: "Between the great man of concepts, Aristotle, and the manners and art of the Hellenes, there stands an immense abyss."[55] She refers to Nietzsche in indicating an abyss not so much between epochs, but more akin to the gulf that separates Heidegger's "whole of beings" and "being in the whole."

In marking a distinction between these two ways of thinking she inverts Rosenzweig's use of the word *Abgrund* (abyss) in referring to the two dangerous voids Goethe was able to bypass, and Nietzsche was not. Kofman suggests that rather than viewing the empty abyss as a threat, one may rather view it as providing the space necessary to open up to a new way of seeing: "The abyss separates but also engulfs; it is thus a metaphor for talking about forgetfulness: about the forgotten import of Presocratic philosophy, of metaphor, and of the totality of instincts as opposed to a single one (i.e., knowledge)." Kofman concludes:

> Nietzsche leaps the empty abyss: he goes beyond the Western philosophical tradition of Aristotle back to Heraclitus, where he takes up the metaphor of world as game or play. He is the disciple who personally re-enacts Presocratic philosophy by reversing the opposition between metaphor and concept, by

reinstating metaphor itself, after its eradication by the concept and within the concept.[56]

Kofman views metaphor in terms that provide room for individual personalities to develop as *fragment* in the face of the whole. In contrast, "the forgetting of metaphor is the eradication of personality." To stick with philosophical concepts maintains the type of totalizing discourse that erases personality.

Kofman's reading of and comments on the notion of abyss in her typological reading of epochs we may also find useful in referring back to Rosenzweig's division between the sinner whose prayer is too late, and the fanatic whose prayer is too soon. For Kofman, it isn't a matter of finding redemption through changing the past (that can never be changed), nor in forcing an impossibly idealistic future. Her reading guards against the urge to return, not through a potentially redemptive mourning, but through an ever-lasting melancholia toward a past utopia we can never reclaim or inhabit. Kofman views Nietzsche as truly living in the right time; his succumbing to metaphoric fancy actually saves him from the either/or dilemma of Rosenzweig's too soon or too late prayer.

Nietzsche writes of the "History of an Error" in Twilight of the Idols, of how "How the 'True World' Finally became a Fable" concerning his own approach to the history of metaphysics. Nietzsche begins his "History of an Error":

> 1. The true world—attainable for the sage, the pious, the virtuous man; he lives in it, *he is it*.(The oldest form of an idea, relatively sensible, simple and persuasive. A circumlocution for the sentence, 'I, Plato, *am* the truth.')
>
> 2. The true world—unattainable for now, but promised for the sage, the pious, the virtuous man ('for the sinner who repents').(Progress of the idea: it becomes more subtle, insidious, incomprehensible—*it becomes female*, it becomes Christian.)
>
> 3. The true world—unattainable, indemonstrable, unpromisable; but the very thought if it—a consolation, an obligation, and imperative.(At bottom, the old sun, but seen through mist and skepticism. The idea has become elusive, pail, Nordic, Königsbergian.)

Nietzsche takes us from post Pre-Socratics in Plato, through the advent of Christianity to Kant. Certainly, as Blondel may agree, Nietzsche is not suggesting an attempt to return to the Socratics or the Presocratics, not in any melancholic and nostalgic return to a previous time or to an existence out of time. Nietzsche continues his history through to the last epoch:

> 4. The true world—unattainable? At any rate, unattained. And being unattained, also unknown. Consequently, not consoling, redeeming, or obligating:

how could something unknown obligate us? (Gray morning. The first yawn of reason. The cockcrow of positivism.)

5. The "true" world—an idea which is no longer good for anything, not even obligating—an idea which has become useless and superfluous—consequently, a refuted idea: let us abolish it! (Bright day; breakfast; return of bon sens and cheerfulness. Plato's embarrassed blush; pandemonium of all free spirits.)

6. The true world—we have abolished. What world has remained? The apparent one perhaps? But no! With the true world we have also abolished the apparent one. (noon; moment of the briefest shadow; end of the longest error; high point of humanity; INCIPIT ZARATHUSTRA.)[57]

Walter Kaufman provides in his note on this passage that INCIPIT ZARATHUSTRA, meaning "Zarathustra begins," functions as "an echo of the conclusion of the 'Gay Science'." Kaufman adds: "Nietzsche had used the first section of the Prologue of Zarathustra, his next work [following *The Gay Science*], as the final aphorism of Book Four, and [gave] it the title: Incipit tragoedia."[58]

Nietzsche's closing, serving as it does to introduce *Zarathustra,* also precedes, in the Appendix to *The Gay Science*, the first poem in his Songs of Prince Vogelfrei titled "To Goethe." The last lines of this poem refer to the end of the second part of Goethe's *Faust*, referencing the "Eternal-feminine" Nietzsche alludes to in Part II of Zarathustra, "On the Poets."[59] Nietzsche describes the poets of his time in terms we may assign to the "protest" rather than dissident poet: "This parable I offer the poets. Verily, their spirit itself is the peacock of peacocks and a sea of vanity! The spirit of the poet craves spectators…"[60] In contrast, ever the dissident poet, Nietzsche's last stanza of his "To Goethe" reads:

World game, the ruling force,
blends false and true:
the eternally fooling force
blends in us too.[61]

Where do we find truth in such metaphorical writing? "The eternally fooling force/blends in us too," as the Eternal-Feminine Nietzsche refers to in Part II of *Zarathustra*, in Faustian longing: "We are covetous…of those things which the old females tell each other in the evening…"[62] Nietzsche identifies this Eternal-Feminine with a secret the grandmother's tell each other, a secret, hidden from view.

PART V: POEM/PLAY TO LIFE/DEATH: "THE DEVIL AND HIS GRANDMOTHER"

Crucifixion
"Do not weep for Me, Mother,
I am in the grave."

1
A choir of angels sang the praises of that momentous hour,
And the heavens dissolved in fire.
To his Father He said: "Why has Thou forsaken me!"
And to his mother: "Oh, do not weep for Me..."

2
Mary Magdalene beat her breast and sobbed,
The beloved disciple turned to stone,
But where the silent Mother stood, there
No one glanced and no one would have dared.
—Anna Akhmatova[63]

Only when we see the star as countenance do we transcend every possibility and simply see. —Franz Rosenzweig[64]

I am—because he saw me: was then face.
Visage before him. If he looked away,
I would be naught. I am—because in me
He shines back to himself. I am—
Reflected. Glow and womb of his conception.
—Lou Andreas-Salomé[65]

For as much as Rosenzweig appears to have dismissed Nietzsche's philosophy for his use of metaphor and poetic imagery, Rosenzweig utilized a very fundamental metaphor in his *Star of Redemption*. He moves from references to death (in terms of philosophical inquiry since Socrates), towards life at the end of the *Star*, inviting his readers to approach the divine, face-to-face in the symbolism (metaphor) of the Star of David as the face of God. The face of God is not literally the Star of David for Rosenzweig, of course; the Star of David likewise doesn't come close to resembling a human face. However, we may find in Rosenzweig's philosophy the use of poetic imagery and metaphor functioning in a way that Nietzsche himself may have blessed.[66] Nietzsche writes in *Birth of Tragedy*: "For a genuine poetry, metaphor is not a rhetorical figure but a vicarious image that he actually beholds in place of a concept."[67] Rosenzweig would arguably agree that the lie of the poet might speak more truth than any totalizing discourse, systematic thinking or overarching analysis.

Rosenzweig's "vicarious image" of the face of God is an actual human face he describes in the image of two overlapping triangles, in reverse fashion, forming the star of David: one triangle points upward, from two cheeks to the forehead, and the other faces downward, from the two eyes to the mouth. The two triangles, in two-dimensional form resemble a star, with the eyes, ears, nose, cheeks and mouth all "standing" for sacred elements of a divine visage. The elements line up as a camera obscura, with the nose at the center point, which Rosenzweig has "belong[ing] to the forehead,' but also "in the sacred (Hebrew) tongue…veritably stand [ing] for the face as a whole."[68]

One may wonder, is there another face of God, one associated not so much with the mystical idealized presence of Shekinah in the Jewish tradition, or a Sophia Wisdom transformed into Holy Spirit in the Christian tradition, but as the very foundation of being, within nature? Can there be a true face of a feminine divine arising from nature? If Nature is a metaphor for a feminine God, how may we view the "vicarious image"?

Feminist philosopher and theologian Grace Jantzen grapples with the issue of how to present a feminine divine within western theology and philosophy based as it is on dualism: "The dualist symbolic with its presumption of domination has been formative of the history and culture of the west"; this is a binary symbolic model based on "the harshness of…mastery [whether of husband over wife, master over slave, employer over employee, humankind over death]." Instead, for a feminist divine Jantzen calls for a religious symbolic whose "patterns would foster respect rather than domination, mutuality in place of mastery." Jantzen notes traditional views of women in relation to nature:

> Women have been conceptually linked in the western symbolic with bodiliness, reproduction, and the material world,…This linkage has been that which stands as the other to masculine self-identity and subtends it. Though the masculine world demeans the female, it is dependent upon it in every way. The material is thus used as a springboard to an ostensibly disembodied transcendence, a climbing up into the head which is in fact possible for men only because their actual bodily needs are being met. Because of the identification in the western symbolic of the female with the material, for women to project a divine horizon, a 'God according to our gender/genre'…it is necessary that this female divine cannot be pure spirit, a disembodied God, but thought of in female terms, just as the disembodied Father God has been thought of in male terms.[69]

How may we open this divine horizon, and in what manner may it most directly and profoundly come forth?

Similar to Jantzen's perspective on history and theology in the west, in her essay "Jesus der Jude," Andreas-Salomé addresses how western divinity came into existence. For Andreas-Salomé, to ask how a man-made divinity becomes "the creative principle of their complete inner life,"[70] is to ask what can be going on in modernity. She avoids the either/or trap of nihilism vs. positivism and instead emphasizes the process of will-defining creative/spiritual genius. Her exploration of religion in her essays "Jesus der Jude" and "Was daraus folgt, daß es nicht die Frau gewesen ist, die den Vater totgeschlagen hat" offer no easy dialectical analysis based on dualistic assumptions.

I find Andreas-Salomé's play "The Devil and His Grandmother" can help envision a maternal divine not restricted through masculinist logic (or in other words, as Jantzen puts it, as "God in a skirt"). I turn now to some aspects of Lou Andreas-Salomé's "The Devil and His Grandmother" in order to envision an approach to a feminine divine associated with nature that could transvaluate binaries and function as one example of Seeburger's aesthetics of nature. There may be ample room for "seeing nature" in terms of a double reading when viewed through the lens which "conflates" nature and the divine in poetic and metaphoric fashion in order to create room for, in the sense of the Aletheia/disclosure/ "reveal"/uncover more, a face of/in a nature through an uniquely identified feminist lens.

Lou Andreas-Salomé wrote "The Devil and his Grandmother" in 1915, after having met Freud, and studying psychoanalysis. It was not published until 1922, and not translated into English until 2016. This work is unique in that it is the only poem/play that Salomé wrote.[71] Playing off the Grimm Fairy Tale by the same name, Andreas-Salomé gives her readers a new myth to help provide insights toward an aesthetics of nature from the ground, or, to utilize a word Andreas-Salomé used in her psychoanalytical writing, *Urgrund,* up.

As I earlier noted, in the introduction to *Sex and Religion: Two Texts of Early Feminist Psychoanalyses*, Matthew Del Nevo and Gary Winship describe the movement in the dialog of Andreas-Salomé's "The Devil and his Grandmother" as "like that of Rosenzweig's *Star of Redemption*," in that they were both were written around the same time, and both move from "from Death to Life."[72] They also note that, "in the dialogue [of the play] the Grandmother is just a voice and not a personal voice but a Voice akin to that of nature, archetypally a Gaia figure," and that Andreas-Salomé writes the Devil as a "post-theological" myth hearkening back to Nietzsche's Zarathustra.

Andreas-Salomé certainly evokes Goethe's *Faust* and in a different manner and extent from Rosenzweig's *Star* and Nietzsche's *Zarathustra*. She rewrote the myth, creating a new one that in many ways follows Rosenzweig's trajectory in the *Star*, not just in moving from death to life, albeit in transvalued fashion, but in showing a face of God through her own imagery/mythology that may not be easy to face. Her play begins in hell, which is the opposite of the face God, and as such, the face of God is replaced with the ass of

Grandmother nature.[73] In an absurd fashion certain boundaries around divinity and nature, or nature as divine, cross over.

The ass of the Grandmother functions as Andreas-Salomé's version of hell as representing death. The life Andreas-Salomé leads her reader to through her narrative is not the Face of God as the star of David, but what has been cut off from God and become abject. Andreas-Salomé's *abject* faceless divine speaks as nature, and when her true face comes forward in one particular scene as a reflection in lake water, it is as the face of an earthly mother expressing compassion, love and sorrow for that which has been cut off from her creation, HER only begotten first-born son, and for the sake/salvation of the afore mentioned father God.

Del Nevo and Winship describe the opening scene of the play emphasizing what is abject:

> At the Beginning of 'The Devil of His Grandmother' the Poor Little Soul is in Hell with the Devil. But this is not the Evil of the Hell of theology and Christian religion. This is abjection. The Devil, Hell, and the Poor Little Soul have abjection in common.

Seeburger's notion of the fragment speaking for the whole comes alive as we read how the Devil carves up his wife on the marital bed, with precise detail; interestingly enough carving a star on her body, then cutting her up into little pieces. These pieces he later throws into the air, and in surrealistic (fetishistic) fashion, they then fly through space and form words for Andreas-Salomé's readers to envision. They speak and/as fragments of the whole, literally, as individual words give meaning to a type of total world/reality. Grandmother as nature, with hell being Grandmother's ass, appears shocking or extreme, too much to take in and handle, too much to face.

Del Nevo and Winship suggest that "Lou's Devil is Zarathustrian. He reels in his godlessness. He simulates God." We see a reminiscence with Zarathustra, but, as Del Nevo and Winship point out, "the way they all revel in their abjection is quite beyond Nietzsche's humourless Zarathustra."[74] Del Nevo and Winship note that Andreas-Salomé's style leans less toward philosophy, even as akin to Nietzsche's philosophy, but more toward Bertolt Brecht. Her poem is set not only as a play, with stage direction, hearkening back to Goethe's *Faust,* but she also includes cinematic elements. Metaphor takes on new meaning; in being a poem, play, and film all at once, her readers experience not only reading a text, but also envisioning action and images. She uses intertitles in her drama; such that the reader's eye must move from text to image to text again.

Andreas-Salomé gives us parody of key Biblical passages, and through the play on words/images, is able to wake her readers to a deeper understanding of life. In transgressing traditional approaches to western theology, particularly Christian theology, the origin of Andreas-Salomé's Devil is actually as *her* first-born son. This Devil has a mother, the same mother in fact as his younger brother, Jesus. Andreas-Salomé presents a unique feminist theodicy here, not only providing a rational for evil in the world, but also actually evoking sympathy for the Devil in rather profound ways. We hear the Devil speak. Both the Devil and Jesus suffered and died, but for Andreas-Salomé, it was the Devil who was the real Savior; it was he who was sacrificed/demonized to save God from being tainted by darkness, so that his children could trust and love him.

> The Voice of the divine feminine, Grandmother nature, explains to the Devil:
> [God] had to do it for his people's sake.
> Human weakness needs God pure and loving.
> That they may trust him fully, and so that
> They dare approach him. So he let you fall.[75]

As the Devil was sacrificed, he, unlike with his younger brother, does so willingly:

> Father! If it Be possible, then may this cup not pass
> Me by, —since you'll not drink it anyhow…
> Father God's eldest son, of noblest birth—
> Nearly still more than God: his most secret
> Most intimate part anxiously renounced.
> The Father stands in shame before his Son:
> 'Dear Father, let it be,' the son tells him,
> If you have blackened me, then I do more:
> I judge myself—that you may not be judged!"[76]

The Devil was the fall guy, who suffers more than his younger brother. Again, it was the Devil who took on the "sin" of God and was scapegoated to save not so much humankind, but God himself.

The Voice as Grandmother nature is the creator of the creator God. The Devil says:

> I understood! Grandmother!—wordless song!
> You sing of God, who impregnated you
> Before you bore him (—in parenthesis:
> A quite godly little maneuver that.)
> Yet, so you must have been there even then,
> And his maneuver does not surpass you.[77]

Grandmother's maternal voice rings in and as nature, providing a transvaluated perspective on becoming abject, fragmented, and whole:

> *Voice [resounding from the mountains].*

You only know the death that you have played,
The death of hell among the disunited,
Who've divided and fragmented themselves.
To enter God is to be perfect wholeness,
Not fragmented: limitless. Not dead: God.
Redeeming you as well would be easy:
Once I unveil my moon, my nightly light
Your little sparks will soon be extinguished,
It will put out what separated you
Deceived and kept you from fullness of being.
Pale moonshine even, merely borrowed light,
Not just the sun, effaces your pretense. [78]

She redeems through the light of her moon that extinguishes his light, not by blowing his light out, but by letting his "glowing embers from the fire-dotted…outline"[79] dissolve into hers. She speaks to him with motherly concern and sympathy:

Voice [resounding from the mountains that stand closest to the Devil].
Poor rascal, you—you're not at fault,—not you!
Simply misunderstood your role, and, thus,
Opposed God, whom you so closely resembled.
The world of matter thus was sin: lustful—
Though that world is motherhood and conception, Is both all and nothing. [80]

Even as the scenes in Andreas-Salomé's play appear surreal, absurd, or at the least farcical, the narrative opens its readers to some heart felt emotion. When the Devil assumes his Grandmother would be more proud of "that grandson, who did redeem the world./While I, the first-born son, in misery/ Here die—forgotten," we read from the "Voice [whispering from the treetops around the rock crevice]/You are not forgotten."[81] The Devil feels her forgiveness—he stands as one who can take in her love: "Listen: this is no wind -- Grandmother's voice!/If she were visible: neither would she,/Appear as a shadow: love would appear." And when she presents herself to him, face-to-face, the Devil says:

But then appeared
To me—oh heavenly! A woman's face.
She bent down close to me, within her hands
The light, from which the darkness fled,—from which
All secret wrongs did fly as well, for in
That countenance of beaming light she wore
The loving gaze that calls one home—[82]

This Divine face does more than just be seen, she looks back—and she does so lovingly. She is not the face of divine retribution or dualisms, of good son and bad son; she loves all. Her face does more than shine, in fact, it weeps:

> *Devil [listening intently; disbelieving, quite].*
> —Grandmother—do you cry—? Is it the wind
> That wails and moans as if autumn had fallen
> All around us? Say: has my tail thus stripped
> The trees—? Just tell me this: could it for me
> For my sake now turn autumn—? Grandmother—
> Do you cry?
> *Wailing Voice.* How could I but cry for my
> Child, who's not my grandson from foreign womb,
> My child, that I did bear in sight of God,
> For you, my son—you, whom I love—
> *Devil [Letting go of the protective wall, throws himself with a joyful cry out into the moonlight].*
> Mother!![83]

The mother's face, as Grandmother nature, expresses sorrow for the loss/sacrifice of her first-born son, the one become abject from the primordial home. The Devil could never ever really play the role of prodigal son, because he is totally "cut off," and can never return to the Father. From the mother, however, he never left. Reading this new Biblical narrative opens up new angles to a familiar story line, in transgressive ways.

Andreas-Salomé expresses quite a redemptive vision of the Voice of Grandmother nature, whose countenance reflects love as nature itself. Near the end of the play, Andreas-Salomé creates somewhat of a radical reversal—with "the end," meaning what should be the entryway into life if following Del Nevo and Winship's reference to a comparison with Rosenzweig's *Star*, but instead Andreas-Salomé's depicts a "scene"/image reflected on the surface of a lake of Mother Nature holding her first-born son as an especially pitiful *Pieta*, in her loving embrace: "In his whole breadth, the lake reflects the picture of a woman holding her dead son on her knees." This image that "was surface," yet "one last depth perhaps" depicting "what came to pass which humans cannot grasp/A secret turned to image which the things/Of day would helplessly misunderstand"[84]

A "feminist" theodicy/theology coming from such a vision of a feminine divine accepting not the Father's only begotten son, but the mothers first born son. Salvation comes not through the crucified birth or resurrection but in the mother's sorrow for her lost son as depicted in the Pieta reflected on the surface of the lake water:

> For this woman with her dark son now holds
> Memorial for him. And everything
> Now joins together in the mother's sorrow.

Do I not first learn of living and of
Dying through him—the son upon her knees?
I thought of life as if it were a mother,
Encircled only by the living as
Her children: thanked her, that myself as well
She called back once again into this sphere.
Her motherhood goes further than I knew:
I realize only now, she motherly
Embraces death as well as one of her
Children—As her dark son—[85]

As opposed to the warning to avoid "the abyss," she shone and spoke through it. In the topsy-turvy narrative, with the play opening with a dream of death, and closing with the true life depicting death, one would presume the message would be that one was never separate from the whole in the first place:

In quiet, I'll await the morning sun,
It shall open all paths for me anew,
That they may guide me. Even if they led
Me endlessly to far, the farthest reaches:
They go nowhere, nowhere away from home.[86]

This feminine divine depicted as nature and divine face accepts not only all of life, but death too. It is as the quizzical notion, contrary to philosophy as form of suicide, or escape from facing our "end," death, of Seeburger's enigmatic statement: "It is only what can never escape its end that can become, in its very ending, unending."[87]

CONCLUSION

In light of some aspects of the fruition of Andreas-Salomé's poem/play in relation to Seeburger's aesthetics of nature, a re-reading of Nietzsche's famous "God is dead" passage may be helpful. Read within the context of a feminine divine as nature perhaps can help more fully to elucidate the need to follow a line of philosophical thinking in the manner of doing, of acting out differently, not only for mere physical survival, but to actually enhance life, ALL life:

All of us are murderers. But how did we do this? How could we drink up the sea? Who gave us the sponge to wipe away the entire horizon? What were we doing when we unchained this earth from its sun? Whither is it moving now? Whither are we moving? Away from the suns? Are we not plunging continually? Backward, sideward, forward, in all directions? Is there still any up or down? Are we not straying as though an infinite nothing? Do we not feel the

> breath of empty space? Has it not become colder? Is not night continually closing in on us? Do we not need to light lanterns in the morning? Do we hear nothing as yet of the noises of the gravediggers who are burying God? Do we smell nothing as yet of the divinc decomposition? Gods, too, decompose. God is dead. God remains dead. And we have killed him.[88]

A radical re-envisioning of nature and a philosophy and theology of nature that can help us deal with our current state of crisis begins with facing that nature is dying. Philosophy can help us face not only life, but also death, in particular this death. Facing this abysmal death can open a pathway for ongoing life, as a "finite open to the infinite." We may begin by writing and reading pure poetry, by being able to sit with the poem/look at the art work as if we were in prayer, that is long enough to let it help us do the work of mourning for us. This may free us from the ongoing Faustian search for something more, in whatever distraction our society "drugged on reproduction" may tempt us with.

Andreas-Salomé offers insights into how an aesthetics of nature works in relation to her philosophical, theological and poetic work; she creates an opening for a feminine face of God, such that through a "conflation" of her art, metaphor and/as life, a way to see beyond the various idols that obstruct our view opens up, so that we may see nature as countenance, and truly see. For, the redemption comes, according to Andreas-Salomé's transvaluated Zarathustra, not through any polemic or war of thoughts, words or weapons, but through the love of life, all aspects of life, including the "Devil," including even death.

BIBLIOGRAPHY

Akhmatova, Anna. "Requiem." *Selected Poems of Anna Akhmatova*. Translated by Judith Hemschemey. Brookline, MA: Zephyr Press, 2000. 147.

Andreas-Salomé. *Looking Back Memoirs*. by Lou Andreas-Salomé. Translated by Breon Mitchell. New York: Marlowe and CO, 1995.

———. Lou, Mein Dank an Freud: Offener Brief an Professor Sigmund Freud zu seinem 75. Geburtstag. Wien: Internationaler Psychoanalytischer Verlag, 1931.

———. "The Devil and His Grandmother" in *Sex and Religion: Two Texts of Early Feminist Psychoanalysis* by Lou Andreas-Salomé, trans. Maike Oergel and Kristine Jennings. New Brunswick: Transaction Publishers, 2016.

———. "Jesus Der Jude," *Neue Deutsche Rundschau* (1896). 342–51.

———. "Was daraus folgt, daß es nicht die Frau gewesen ist, die den Vater totgeschlagen hat" Almanach des Internationalen Psychoanalytischen Verlags: Wien (1928): 25–30.

Blondel, Eric. "Nietzsche as Metaphor," in *New Nietzsche: Contemporary Styles of Interpretation*, edited by David B. Allison Cambridge, MA: 1995. 150–175.

Goethe, Johann Wolfgang von. *Faust: A Tragedy*. Translated by Walter Arendt. New York: WW Norton 2001.

Heidegger, Martin. "Origins of the Work of Art," in *Basic Writings*. edited by David Farrell Krell. San Francisco: Harper Collins, 1993. 143–212.

Jantzen, Grace M. *Becoming Divine: Towards a Feminist Philosophy of Religion*. Bloomington, IN: Indiana University Press, 1999.

Kofman, Sarah. "Metaphor, Symbol and Metamorphosis," in *New Nietzsche: Contemporary Styles of Interpretation*, edited by David B. Allison Cambridge, MA: 1995. 201–214.

Nietzsche, Friedrich. *Gay Science.* Translated by Walter Kaufmann. New York: Viking Press, 1974.

———. *Thus Spoke Zarathustra* in *The Portable Nietzsche.* Edited by Walter Kaufman. New York: Viking Press, 1972. 112–439.

———. "Twilight of an Error," in *Twilight of the Idols*, by Friedrich Nietzsche. Translated by R. J. Hollingdale. London: Penquin Classics, 1990, 120.

Rich, Adrienne. *What is Found There: Notebooks on Poetry and Politics.* WW Norton & Co: New York, 2003.

Rosenzweig, Franz. *The Star of Redemption.* Translated by Barbara E. Galli. Madison, WI: University of Wisconsin Press, 2005.

Rosenzweig, Franz. *The Star of Redemption.* Translated by William W. Hallo. Notre Dame, IN: University of Notre Dame Press, 1985.

Seeburger, Frank. "Seeing Nature Whole in Fragments." 2017. In Press.

Winterson, Jeanette. *Art Objects: Essays on Ecstasy and Effrontery.* New York: Vintage Books, 1995.

NOTES

1. Friedrich Nietzsche, "Twilight of an Error," from *Twilight of the Idols*, by Friedrich Nietzsche, trans. R. J. Hollingdale (London: Penquin Classics, 1990), 120.

2. Lou Andreas-Salomé, *My Thanks to Freud* (*Mein Dank an Freud*): Open Letter to Professor Sigmund Freud for His 75th Birthday (1931), trans. William Needham, https://archive.org/details/MeinDank. ["Menschleben—ach! Leben überhaupt—ist Dichtung. Uns selber unbewusst leben *wir* es, Tag um Tag wie Stück um Stück—in seiner unantastbaren Ganzheit aber lebt es, dichtet es *uns*. Weit weitab von der alten Phrase vom "Sich-das-Leben-zum-Kunstwerk-machen"; wir sind nicht unser Kunstwerk." *Mein Dank an Freud: offener Brief an Professor zu seinem 75. Geburtstag* (14). Wien: Internationaler Psychoanalystischer Verlag, 1931.] Also in epigraph of *Looking Back Memoirs*, by Lou Andreas-Salomé. Trans. Breon Mitchell (New York: Marlowe and CO, 1995).

3. Johann Wolfgang von Goethe, *Faust: A Tragedy* ("Ich bin ein Teil des Teils, der einstmal alles war"), *Faust,* 1349 trans. Walter Arendt. (New York: WW Norton 2001), 37.

4. Rosemary Radford Ruether offers a critique of feminist theologians claiming "there was once a culture...possibly worldwide, for most of human history, until the last few thousand years—in which a matricentric, if not matriarchal, society flourished. Humans were in harmony with one another and nature, and a female-personified deity expressed the immanent life energy that cycled through the earth as one community." This is a past, Reuther asserts, that cannot be verified empirically. Any future society based on it would also be unrealistic and impractical, if not impossible. See Rosemary Radford Ruether, *Goddesses and the Divine Feminine: A Western Religious History* (University of California Press: Berkeley, 2005).

5. Franz Rosenzweig, *The Star of Redemption,* trans. William W. Hallo (Notre Dame, IN: University of Notre Dame Press, 1985), 286.

6. In their introduction to *Sex and Religion: Two Texts of Early Feminist Psychoanalysis*, Matthew Del Nevo and Gary Winship suggest a comparison between Rosenzweig's *Star* and Andreas-Salomé's "The Devil and His Grandmother": "The movement of the dialogue [in "The Devil and His Grandmother"] is like that of Rosenzweig's *Star of Redemption* (written around the same time) from Death to Life." See *Sex and Religion: Two Texts of Early Feminist Psychoanalysis* by Lou Andreas-Salomé, trans. Maike Oergel and Kristine Jennings. (New Brunswick: Transaction Publishers, 2016), xvi.

7. Frank Seeburger, "Seeing Nature Whole in Fragments" 2017. In Press.

8. Franz Rosenzweig, *The Star of Redemption*, trans. Barbara E. Galli (Madison, WI: University of Wisconsin Press, 2005), 15.

9. Certainly Nietzsche delved into a type of struggle between the Apollonian form and Dionysian creative energy in exploring the power of art, specifically in his early work *The Birth*

of Tragedy, but also throughout his life work, albeit in ever more subtle and expanding nuance. This is particularly evident in *Thus Spoke Zarathustra.*

10. Martin Heidegger, "Origins of the Work of Art," in *Basic Writings*, ed. By David Farrell Krell (San Francisco: Harper Collins, 1993), 184.

11. Adrienne Rich, *What is Found There: Notebooks on Poetry and Politics* (WW Norton & Co: New York, 2003) xiv. Rich here refers to the Walter Benjamin's "angel of history" from Benjamin's oft-quoted commentary on Paul Klee's painting: "This is how one pictures the angel of history. His face is turned toward the past. Where we perceive a chain of events, he sees one single catastrophe which keeps piling wreckage upon wreckage and hurls it in front of his feet. The angel would like to stay, awaken the dead, and make whole what has been smashed" by the storm of progress." Walter Benjamin, "Theses on the Philosophy History" in *Illuminations*, ed. by Hannah Arendt. New York: Schocken Books, 1968 (p. 257).

12. Rosenzweig, trans. Galli, 15-16.

13. Seeburger, "Seeing Nature Whole in Fragments" 2017. In Press.

14. Seeburger, "Seeing Nature Whole in Fragments" 2017. In Press. For recent analysis of the Greek *agon* in relation to politics and Nietzsche's thought, see Yunus Tuncel, *Agon in Nietzsche* Marquette University Press: Milwaukee, WI, 2013) and Christa Davis Acampora *Contesting Nietzsche* (University of Chicago Press, Chicago and London, 2013).

15. Seeburger, "Seeing Nature Whole in Fragments" 2017. In Press.

16. Rosenzweig, trans. Hallo, 4. ["Es ist wohl notig, dab der Mensch einmal in seinem Leben heraustrete; er mub einmal die kostbare Phiole voll Andacht herunterholen; ermub sick einmal in seiner furchtbaren Armut, Eisamkeit und Losgerissenheit von aller Welt gefuhlt haben und eine Nacht lang Aug in Auge mit dem Nichts gestanden sein. Aber die Erde verlangt ihn wieder." Franz Rosenzweig, *Der Stern Der Erlösung,* intro. Reinhold Mayer (The Hague, Netherlands: Martinus Nijhoff, 1976), 4].

17. In his translation of *The Star of Redemption*, William Hallo references *Faust* directly in this passage. Barbara Galli, in her translation, does not. See Barbara Galli, *The Star of Redemption* (Madison, WI: University of Wisconsin Press, 2005), 10. For the reference to Faust see Johann Wolfgang von Goethe, *Faust: A Tragedy,* 733 trans. by Walter Arendt. (New York: WW Norton, 2001), 21.

18. Rosenzweig, trans. Hallo, 4.

19. Rosenzweig, trans. Hallo, 3.

20. Jeanette Winterson, *Art Objects: Essays on Ecstasy and Effrontery* (New York: Vintage Books, 1995), p. 12.

21. Winterson, 19.

22. Winterson, 8.

23. Winterson, 8.

24. Rosenzweig, trans. Galli, 16.. ["Der Mensch in der schlechthinnigen Einzelheit seines Eigenwesens, in seinem durch Vor- und Zunamen festgelegten Sein, trat aus der Welt, die sich als die denkbare wußte, dem All der Philosophie heraus," Mayer, 10]

25. Nietzsche refers to *ressentiment* as acting out of revenge: "For that man be delivered from revenge, that is for me the bridge to the highest hope, and a rainbow after long storms" ("On the Tarantulas" in *Thus Spoke Zarathustra*) in *Portable Nietzsche*, trans. By Walter Kaufmann (New York: Viking Press, 1968), 211. See also *Genealogy of Morals*, section 10-12.

26. Quoted in Adrienne Rich, *What is Found There: Notebooks on Poetry and Politics* (WW Norton & Co: New York, 2003), xiv-xv. Even as Rosenzweig may not have delved into the ramifications of Nietzsche's insights in relation to environmental injustices, Rosenzweig did present Nietzsche as one able to function as a signpost toward a clearer understanding of our relationship with nature and/as other.

27. Rosenzweig, trans. Hallo, 10.

28. Adrian Del Caro acknowledges Nietzsche's influence on Thomas Mann in terms of art and politics, asserting that "Mann had a strong, almost irrational tendency to ascribe his own…romantically derived views to Nietzsche, as if the latter were the actual…romantic influence on him," such that "when it came time to finally and definitely break with Germans in political terms, and only in the aftermath of World War II, Mann felt that he had to break with Nietzsche as well, and in doing so he branded Nietzsche a hopeless Romantic, and a hopeless

aesthete, and laid partial blame for Germany's fascist excess at Nietzsche's feet." Adrian Del Caro, *Grounding the Nietzsche Rhetoric of Earth* (Berlin: Walter de Gruyter, 2004), 426.

29. Rosenzweig, trans. Galli, 15.

30. In his essay "Rosenzweig Versus Nietzsche," Richard Cohen provides a detailed approach to the ambiguity in Rosenzweig's approach to Nietzsche. See Richard A. Cohen, *Elevations: The Height of the Good in Rosenzweig and Levinas* (Chicago: University of Chicago Press, 1994), 346-66.

31. Rosenzweig, trans. Hallo, 9. Cohen points out that Rosenzweig refers to Nietzsche "at the book's three pivotal points, in the introduction to the three Parts." Whereas he asserts that "Nietzsche's first appearance…is positive, his second appearance is both positive and negative, and his third appearance is negative, " I find there to be ambiguity even in the first appearance, perhaps hinting at Rosenzweig's more negative critique of Nietzsche in the introduction to the Part III. Cohen, "Rosenzweig Versus Nietzsche," 347.

32. Rosenzweig, trans. Galli, 304. Rosenzweig critiqued two phases in the history of Christianity, the first he called the Petrine, the second Pauline, each with inherent dichotomies. See *Star*, trans. Hallo, 278-286 and trans. Galli, 296-304.

33. Both Barbara Galli and William Hallo reference this quote to Mephistopheles (*Ich bin ein Teil des Teils, der einstmal alles war)*. Johann Wolfgang von Goethe, *Faust: A Tragedy* trans. Walter Arendt (New York: WW Norton, 2001), 37.

34. Rosenzweig, trans. Hallo, 285-86.

35. Rosenzweig, trans. Galli, 286.

36. Rosenzweig, trans. Galli, 286.

37. Rosenzweig, trans. Hallo, 275.

38. Rosenzweig, trans. Hallo, 287

39. Rosenzweig, trans. Galli, 304.

40. Rosenzweig, trans. Hallo, 274.

41. Rosenzweig, trans. Hallo, 274-75.

42. Rosenzweig, trans. Hallo, 275.

43. Rosenzweig, trans. Hallo, 286.

44. Rosenzweig, trans. Hallo, 271.

45. Eric Blondel, "Nietzsche as Metaphor," in *New Nietzsche: Contemporary Styles of Interpretation*, ed. David B. Allison (Cambridge, MA: 1995), 150.

46. For example, Jacques Derrida, Gilles Deleuze, Paul de Man, Michel Foucault, Simone De Beauvoir, Martha Nussbaum, and Paul Ricoeur.

47. Kofman, 209.

48. Blondel, 150.

49. Blondel, 171.

50. Seeburger, "Seeing Nature Whole in Fragments" 2017. In Press.

51. Kofman, 210.

52. Kofman, 212.

53. Kofman, 210.

54. Kofman, 212.

55. Kofman, 210 [Nietzsche, *PB*, "The Last Philosopher," § 61].

56. Kofman, 211.

57. Nietzsche, "Twilight of the Idols, *The Portable Nietzsche* (New York: Viking Press, 1972), 485-486.

58. Walter Kaufman, in *The Portable Nietzsche* (New York: Viking Press, 1972), 486.

59. Kaufmann notes the verses from Goethe's *Faust*: "CHORUS MYSTICUS: What is destructible/Is but a parable;/What falls ineluctably/The undeclarable,/Here it was seen,/Here it was action;/The Eternal Feminine/Lures to perfection." Kaufmann, *The Portable Nietzsche*, 193-194. See also Johann Wolfgang von Goethe, *Faust: Part One and Sections from Part Two*, trans. Walter Kaufmann (Garden City, NY: Doubleday & Co, 1961), 503.

60. See Friedrich Nietzsche, *Thus Spoke Zarathustra*, "On Poets," Pt II, in *The Portable Nietzsche* trans. Walter Kaufmann (New York: Viking Press, 1972), 238-241.

61. Friedrich Nietzsche, *Gay Science*, trans. Walter Kaufmann (New York: Viking Press, 1974), 351.

62. Nietzsche, "On Poets," 238.

63. Anna Akhmatova, "Requiem," *Selected Poems of Anna Akhmatova*, trans. Judith Hemschemey (Brookline, MA: Zephyr Press, 2000), 147.

64. Rosenzweig, trans. Hallo, 422.

65. Lou Andreas-Salomé, "The Devil and his Grandmother" in *Sex and Religion: Two Texts of Early Feminist Psychoanalysis* trans. Maike Oergel and Kristine Jennings (New Brunswick: Transaction Publishers, 2016), 74.

66. Even though Nietzsche wasn't able to respond to Rosenzweig personally, perhaps if he could have, there would have been some loving struggle, agon, even love between the two as between friends.

67. Blondel, 150. Blondel quotes Nietzsche, "The Birth of Tragedy" in *The Birth of Tragedy and the Case of Wagner*, trans. Walter Kaufmann (New York: Vintage Books, 1967), 63.

68. Rosenzweig, trans. Hallo, 423.

69. Grace M. Jantzen *Becoming Divine: Towards a Feminist Philosophy of Religion* (Bloomington, ID: Indiana University Press, 1999), 269.

70. Lou Andreas-Salomé, "Jesus Der Jude," *Neue Deutsche Rundschau* (1896), 342–51.

71. Ernst Pfeiffer, in his Bibliographical Notes to Andreas-Salomé's Memoirs he describes "Der Teufel und seine Gorssmutter" as a "Dream play." See *Looking Back Memoirs*, by Lou Andreas-Salomé. Trans. Breon Mitchell (New York: Marlowe and CO, 1995), 138.

72. Del Nevo and Winship, xvi.

73. Andreas-Salomé, "The Devil and His Grandmother" lines 91-92, 41. The Devil says to the Poor Little Soul: "It's just an ass against which you all sit./Grandmother's backside is what you have chosen." Andreas-Salomé juxtaposes hell not with God, or the face of God, but with Abraham's lap. See line 96.

74. Matthew Del Nevo and Gary Winship, xxi.

75. Andreas-Salomé, "The Devil and His Grandmother" lines 968-71, 75.

76. Andreas-Salomé, "The Devil and His Grandmother" lines 1035-37, 76-77.

77. Andreas-Salomé, "The Devil and His Grandmother" lines 938-42, 74.

78. Andreas-Salomé, "The Devil and His Grandmother" line 999-1011, 76.

79. Andreas-Salomé, "The Devil and His Grandmother" lines 14-15, 39.

80. Andreas-Salomé, "The Devil and His Grandmother" lines 955-960, 74.

81. Andreas-Salomé, "The Devil and His Grandmother" lines 1097-1100, 80-81.

82. Andreas-Salomé, "The Devil and His Grandmother" lines 118-1122, 81.

83. Andreas-Salomé, "The Devil and His Grandmother" lines 1154-60, 83.

84. Andreas-Salomé, "The Devil and His Grandmother" lines 1206-15, 85.

85. Andreas-Salomé, "The Devil and His Grandmother" lines 1226-38, 86.

86. Andreas-Salomé, "The Devil and His Grandmother" lines 1250-55, 86.

87. Frank Seeburger, "Seeing Nature Whole in Fragments" 2017. In Press.

88. Friedrich Nietzsche, *The Gay Science* trans. Walter Kaufmann (New York: Vintage Books, 1974), 181.

Chapter Six

Towards a Cathartic Ecology

J. Porter and Joshua Coleman

> These earliest monastics practiced a purification of the soul that correlated to the purity of the place, a process they called Katharsis. This purification, in turn, allowed them to see divine realities everywhere, and this mystical seeing they called *theoria* or contemplation… seeing the divine within all things. —Bruce Foltz, *The Noetics of Nature*

> Beauty is unbearable, drives us to despair, offering us for the minute the glimpse of eternity that we should like to stretch over the whole of time. —Albert Camus, *Carnets: 1935–1942*

BEAUTY AS CATHARSIS

Our collective responses to the current ecological crises mirror many common responses to trauma wherein suffering cannot entirely pass even if some recovery ensues. This means that the trauma cannot be transfigured as mere ordeal, "in itself," but involves the exhaustion of time as one of its resources of processing. Critically, however, as with this crisis, which is not merely an ecological crisis, mirrored are those traumas that are self-induced. At our own peril, we have responded to such indelible habitat loss by becoming more deeply entrenched in our most damaging moral and material habits. Our general answer has been to abdicate both reality and responsibility and bury our heads deeper. Some remain in complete denial (authentic or performed), doing their best to politicize the issue and obfuscate all dispatch of life dying so unreservedly around us, clearly not wanting to know what the science reveals.[1] Whatever the response, no one is spared from such collective trauma and, perhaps most often, the individual vacillates between denial and helplessness, afraid of the overbearing guilt that might ensue alongside the daunting task of calling into question the values by which we've lived to this point.[2] While collective trauma is timeless, unpassing in its passing, in this crisis it arrives space-less. Without material place, within climate models, free of the human imbrue, we gather now in nowhere; we gather in the digital imaginarium in a self-induced blindness where we see but do not sense; where the hastening flow of ghosts, of flora and fauna, released from their final sky come to live with us in some edenic digital wilderness. Even if the real disappears, we have learned to be satisfied with the copy, not terribly concerned about the difference.

Fyodor Dostoevsky famously said, "Beauty will save the world," a comment easily dismissed as a romantic one-off. But if he (whose life was spared moments before death by firing squad) was not being romantic, it seems a discussion of what he meant by beauty is warranted. As will be clear, the religious context of this quote is not superfluous, as the Byzantine Christian logic to iconography, one which provides a way of perceiving beauty as a relation through an unsettled image, to a past and future at once, has everything to do with our ability to experience past trauma within the current crisis.[3] This ability to "let trauma traumatize" (in the words of Frank Seeburger), as part of our process of moving forward in the ecological discussion, gains meaning when beauty is seen at the heart of our collective trauma; not simply its progressive extinction in our age, but that its very experience is itself traumatic and threatening, a reality which leads to a variety of responses, many of which hasten the demise of the earth. The ability to withstand the experience of beauty as a wound is central to moving forward to the subsequent reception of beauty as a salvific force, one that changes our relation to nature from within.

To be clear, the sort of knowledge of the earth had by way of empirical observation has obvious import, for without the scientific approach to the natural world we may not be aware of the human contribution to its ongoing changes; melting ice caps, increasing disappearance of animal species, depleting ozone, just to name a few. But if this approach, alongside value assigned to portions of the earth by the market economy, are the *only* ways we relate to the natural world, there are critical truths being ignored which would more immediately impact the human/nature relationship toward a re-ordered future, despite irreparable damage which has already occurred. The degree to which we can admit scientific truths about the current environment has everything to do with other approaches to truth pertaining to Psychology, Philosophy and Theology. Contemporary British thinker Philip Sherrard writes:

> Our perception of a tree, a mountain, a face, an animal, or a bird is a reflection of our idea of who we think we are. What we experience in these things is not so much the reality or nature of these things in themselves as

> simply what our own limitations, spiritual, psychological, and physical, permit us to experience of them. Our capacity to perceive and experience is stereotyped according to how we have molded our own image and likeness.
>
> That means that before we can deal with the ecological problem, we have to change our world image, and this in its turn means we have to change our self-image. Unless our own evaluation of ourselves and what constitutes the true nature of our being changes, the way we treat the world about us will not change either.[4]

Earlier in the same article he states plainly, "It is first of all a crisis concerning the way we think. We are treating the planet in an inhuman, god-forsaken manner because we see things in an inhuman, god-forsaken way. And we see things in this way because that basically is how we see ourselves."[5]

This last admission is crucial, as the trauma which comes through such confession is the only way forward, the only way we will alter our perception to reality in meaningful ways that will in turn alter our practices. Without this admission, the trauma of past destruction tends to multiply at increasing rates, one after another, as we cover the last pain as soon as the next arises, desperately holding on to ideology to shade us from what we have done and are doing and, as Sherrard suggests, devaluing our very selves in this process. Such admission awakens us to our evidential enclosure within vanishing life, of species upon species fading before their own fair time; and bare life deepening in muteness, not silence, around the missing. Without clarity in terms of what is in fact seen and heard within *and by* nature, we are in grave danger of resigning ourselves, not only to the earth as commodified object, but to a disappearance of Life which mimics the disappearance of the political realm into the economic.[6]

This overtake of the political by the economic has great import for our purposes, as it is no radical suggestion that the economic value of the earth operates the current politics and policies in charge of American environmental protection. Once one relinquishes the moral sense of the living world, it is the easiest metaphysical act to change from, "Nature is good" to "Nature is a good." Easier still is to then let separate political Institutions decide value or even multi-use value, according to market assessments as the goods enter the mercantile logic of profit maximization. So the difficulty of admission is multilayered, made more difficult by the need to unravel political ideologies as having economic motive before even looking closely at the now common way of approaching the earth as commodity.[7] Ideologies are so entrenched that policies are now intentionally constructed to look the other way, incorporating science inasmuch as it befits short-term profit with seemingly little interest in a current common good, much less a future one. The earth, the place we hold universally as common dwelling *and* common resource, the Biblical place of theophany as God's dwelling, is not viewed as part of a common good but rather as a market commodity for which we are to compete, win or lose.

Another difficulty in admitting ecological offenses lies in the fact that we have to include methods other than purely empirical ones to evaluate them, as science, at least that which is commonly seen as "hard science," is not endowed with the means to admit trauma, the lingering aftermath of violence beyond what is empirical, and yet which appears time and again as a driving force of history. Empirical observation may recognize certain behaviors that suggest trauma, but to reduce trauma to those behaviors would only perpetuate the trauma, looking at symptoms and ignoring root causes. Foltz writes:

> My conclusion, then, is simply that the ontological demise, which has long been underway, has been masked and concealed by the aesthetico-theological demise in modern thought and sensibility, under whose influence we have learned to encounter nature as if it were, in fact, already dead. And to the extent that environmental philosophy bases itself exclusively, or even primarily, upon the discursive rationality of the sciences—which by their very methodology deliver nature as self-subsistent, self-enclosed system—rather than upon the mythico-poetic vision, which approaches nature iconically, as issuing from a distance and invisibility that enchants it and animates it (environmental philosophy is complicitous in this ontological demise) precisely through inadvertently concealing it. [8]

Admitting such "soft science" concerns as trauma may run counter to the approach which says we have to look at what we can all agree to, as the approach to knowledge found in the natural sciences is typically proposed to be objective; an objectivity which not only requires distance by the observer (literal, emotional and existential), but that the personality of the observer be removed for the objectivity to remain. With some places of overlap, the knowledge of the Humanities, especially for Existential Philosophy and Theology, seems to be the opposite, that to know a person or persons requires intimacy and the collapse of distance, that there is no observer and observed, but all parties and their personalities are critical to the equation that leads to relational knowing or what Gabriel Marcel called intersubjectivity. The latter conception implies a communion of perspectives and persons, not a melding of essences. But even here we have to look more closely at the problem of subjectivity and when it actually arrives with regard to what it has experienced.

In other words, it follows that this latter knowledge takes time, that it is crucial to wait to get the full picture of the situation, as the warming of the climate is the physical, "behavioral response" to the traumatic practices, but this is only part of the overall picture insofar as nature is meant to do more than be there for use. If Sherrard is right that our low value of nature is an extension of our low value of ourselves, and if most would not rest with evaluating ourselves through strictly medical or biological terms, then a certain psychological approach to our relation to nature (especially to the denial of the current crisis) proves vital to uncover the varied roots of the issues.

Two crucial and related differences in these approaches to truth have to do with the quality of relationship and with time, or the time it takes for such relational truth to reveal. For obvious reasons, each approach to truth mentioned above will also have dramatically differing conceptions of beauty. If beauty is not primarily a scientific concept or is not only for the perpetuation of a species or some other utility, and if it requires this sort of intimacy and time to even know what it is, then it follows that beauty is (or can be) a relation of perspectives, a two-way street of giving and receiving. If this is true, it cannot be had by projected desires of objectification for use. In fact, such methods preclude access to beauty as a mode of relation.

Again, one approach to knowledge advertises objectivity by way of invulnerability, yet the objectivity is often taken at weighed expense of the "object" in the name of inquiry.[9] In other words, this sort of universal or categorical knowledge often precludes the knowledge of particulars by doing away with its methods of intentional intimacy, of desire and longing, and of a truth which comes through waiting in suffering vulnerability.

With regard to the natural world as commodity, inasmuch as nature is used for a purpose or as object, the "truth" of its value is a quick reveal. The sort of knowledge sought in this case is had at "face value," and such value is determined by how it compares to other points of nature for similar use. In short, they are known by comparison for use, not in and of themselves, much less by way of an intimate indwelling. In this way the object of the scientific method and the commodity of capitalism is purged from its inscribed environ without thought of how that ingresses and effects its truth. Moreover, it is a conscious truth (as opposed to unconscious truths which will be explored in the next section), not asking the one relating to nature to alter her own self-relation to know nature.

But again, insofar as we are in relation to nature for its own sake, the truth revealed by that relation comes slowly, and yet the market does not wait. The megafarm holds no Sabbath for the land as prescribed in the Hebrew Scriptures. Farm one crop until the land is barren and move on, there are shareholders to whom we are now beholden. As such, the truth of nature is not given the time to be seen beyond its instrumental status.

The attempt to possess simultaneously denies a two-way relation. In such suppressive denial, the possessors of nature become themselves possessed; in this Anthropocene age, man finally tautologizes himself, and has left his baleful footprint within the center of the very weather that can destroy him. The poisonous traces of our making, in the emissions we see and fail to remember and deny, have changed the sky into something further than can be forgotten, and will at times descend like a threshold god to unmake us.

By turning toward our self-induced ecological trauma, we would less likely be attracted to continuing the same destructive path. The alcoholic continues drinking because sobriety is too daunting a task, the heroin addict might not be able to fathom the inevitable and excruciating pain of withdrawal. Our past and present orientation to nature are the first and most critical battle lines, and our ability to wait for truth (both good and traumatic) and beauty as relational concepts within nature are central to meaningful change.

This is where seeing *iconically,* rather than only empirically or analytically, is so important. Just as getting to know a person takes time, time to get "a sense" of who they are, a "sense of humor," etc., the approach to iconography in the Christian East works in a variety of directions. While intentionally *not* depicting a realistic present, the viewer actively pieces together a window to the past and future, bringing past and future into the present, the "fullness of time." This makes possible to glimpse what was, not literally or in false mimesis, but to "get a sense" of it and whatever that sense brings. In the case of self-induced trauma, the sense brought is typically sorrow, sadness and a desire to amend one's actions; but it is also typically brought slowly, as one's self-critique over such trauma is in some ways more complicated in its reveal than a trauma brought from someone else. Thus "letting trauma traumatize" is necessary not only to know what to do moving through the current dilemma, but as important, to *desire* what to do.

The realism of the Renaissance approach to visual art (as just one example), in its attempt to solidify the present as empirical and literal, does not abide past trauma or glimpse the future path. It does not act as a window to and fro, as it presumes to present "what is," implying that is *all* there is and that the viewer has no role but to receive it.[10] The viewer is not asked to piece anything together, be it past or future, but passively accepts "the real" as what is presented to the conscious present, or what the conscious is able to presently receive. Such beauty, in this case, is truly "skin deep." By seeing iconically as (at least) a two-way relation, nature reveals its own moral order as relational, without the imposition of a constructed institution, capitalist or other, which places nature within an abstract, narrowly defined value system for other ends than the relation to nature for its own sake.

This iconic relationship to nature and time better enables us to permit the experience of a collective and self-induced trauma, to accept the pain of letting beauty die before moving forward; or, more often, to realize we have already let much beauty die and to allow that reality a realm of disclosure. To wait for catharsis is to fully experience the degradation, a degrading which seems too much to bear but which is necessary in order to be changed. To avoid such a process is to resist the internal change which would reorient us moving forward. More specifically, in this case, true catharsis has to include an altered understanding of beauty as both traumatic and healing.

In his work, *The Cost of Discipleship,* Lutheran Pastor and Nazi resister Dietrich Bonhoeffer wrote, "Cheap grace is the preaching of forgiveness without requiring repentance, baptism without Church discipline. Cheap grace is grace without discipleship, grace without the cross, grace without Jesus Christ"[11] And so it is with "cheap catharsis." Of that which Theology can add to the discussion, most immediate is the proposal that we can see nature as having depths of sacred dwelling, well beyond the surface or as instrument for profit; but, critically, this perception comes through an ascetic and difficult path of *metanoia* or repentance. To think we can "fix" the earth's problems without an ontological change towards it is the same hubris of a lover who has hurt the beloved, asked to be back in close relationship, and yet has not felt the trauma of what he did to break the communion in the first place. Even if that person does not repeat the specific behavior, he is not changed through the repentance of his reorientation to being and the beloved, continuing the orientation which led to the communal destruction. In this sense, the communion remains broken and unhealed. Moreover, the attraction to the destructive behavior likely remains. Analogous to Bonhoeffer's "cheap grace," this attempt to "fix" past destruction of the earth without admitting its trauma can be seen as "cheap catharsis." More specifically, a cheap catharsis refuses the responsibility and trauma of past eroded beauty, unable as it is to engage a necessary eros with regard to nature, or that love which forces us outside of ourselves in pure vulnerability.

The incorporation of eros toward nature moves us past words like stewardship, a term which implies a sort of "keeping" of nature as object, regardless of our relation to it. Eros reveals and conceals, and in so doing requires a two-way relationship. If truth and beauty can be seen as relations and not essential concepts, then caring for nature is an ongoing, changing relation, one which, by refusing the objectification that comes with "keeping" nature, forces us "erotically" outside ourselves. Such orientation at the same time reveals our own interiority, beyond the cultural categories of identity, wherein there is no bifurcation or bedrock split between nature and culture, as neither nature or culture operates as objective self-subsistent systems. Indeed, it is nature which keeps us in an erotic exchange that issues first from outside us in its "inestimable, unrepeatable, breathtaking givenness that announces new life."[12] We become liberated, in such erotic address, from a reductive identity that disavows all possibility of recollecting oneself within a greater "we" that defies pure anthropocentrism. Eros, in its keeping of us, creates a desire that is foreign, almost inimical, to our consciousness, requiring further exodus to meet in the place where the beloved might further reveal itself as open to our care. Beauty is then a trauma that forces us to leave home, the leaving itself central to a healing future relation.

EROS AS EXODUS

Conceiving of nature as self-sufficient and distinct from us is actually (perhaps ironically) a way of making nature into an idol of our pleasure, a meeting place of non-discovery, where science and commodification can plunder unobstructed for their own utility. In relation to the idol, there is no taking leave, no exodus from home, because there is no desire that wounds so deeply as to call itself love. There is no vulnerability on the part of the idol maker, who knows better than anyone the idol cannot reveal anything.

Eros is a longing, incurred at first from the outside where the beauty of the given (ness) burrows into us, despite our constitutive attempts to thwart such intrusion; *eros* demands a journeying beyond oneself to the purity of place that no one can find alone. Such recollection of self beyond the purely self-enclosed parameter is an exodus, wherein true repentance can occur because the systematic categories of cheap catharsis have been exchanged for a turning, a metanoia and desire to practice ecological care. The journey requires a prior emergence, where desire creates the erotic action of exodus from the ontological vision, in the practice of ecological care— if it is to be care at all. Such discerning of ourselves within ecological care once again restores to us the meeting place where we are able to open in tenderness to the wound we have both induced and inherited to obfuscate a longing to belong to a natural world we deem already dead.

It is now cliche to say that one must "go back to go forward" to find a degree of liberation within trauma, and yet health and love belong to a process far more multidirectional than back and forth.[13] It might be said that one must go backward and forward to "be here now," the back and forth often leaving one a dark present, a present where waiting allows such non-linear coming and going and piecing together, the imagination playing a crucial role in this process. However, even admitting this seems woefully insufficient to capture or universalize the complexity involved for liberation within a *particular* trauma.

Here we take cues from the Hebrew Book of Exodus as a kind of liberation within trauma. As a sort of "erotic journey," there is seemingly no point to their wanderings. There is the assurance of a "promised land," but in the meantime there seems to be no direct route to get there. They are asked to leave their household gods, ones which obtain to little more than a simulacrum of themselves, step outside themselves in pure vulnerability, and follow a God beyond all imaging. The temptation to find a "point" to all of this may be exactly the logic which gets in the way of the transformation of being. In truth, even pointing *this* out may get in the way. And because *eros* is a longing for what in part one does *not* have, there is a "possessed" violence that leads to possessing the Promised Land. You could argue that this violence is the fall from grace and founding of the state, the entity for which violence is taken up to protect it as one's own possession.

Regardless, during various times in the desert, the waiting for God is cut short in frustration, the golden calf representing, among other things, access to an immediacy and a totality not available from the God they have been following to this point. It also suggests that "what you see is what you get," as the calf is dead and has nothing to offer back. It is significant that they refer to the idol as the god who took them out of Egypt, and not simply a different god, maybe one of the Egyptian gods. It seems they want Yaweh to be different, predictable and controllable. As with the temptation to use and control the earth itself, the Hebrews attempted a "dominion" over God, the two-way relation now reduced to one way. The relation to Yaweh is no longer to Yaweh (despite the same name and intention), but to an attempted copy. By relating to nature as object, we are not relating to nature at all, but to an idea in its place. As a one way relation, the one relating remains "possessed" by that idea, invulnerable and unable or unwilling to wait and receive *from nature*.

It is precisely through a *lack* of totality which allows the simultaneous experience of *what is and of what is not*, as well as for sustaining the "erotic" desire for what is not yet disclosed. To use Martin Buber, this is central to an "I-Thou" relation, the relation maintaining its depth by way of a co-inherence of revealing and concealing, that there is always more to know. The I-Thou becomes an "I-It" the moment it is believed to be had in full, the moment it is believed there is no more to know. This is when *eros* dies and catharsis ends.

The experience of catharsis, as with the experience of *eros*, is allowed to arise insofar as we forgo the ideality (or idolatry) of having the whole. Whether it is the illusion of the whole, literal access to the past traumatic event, or that of possessing the whole person in erotic relation, all that *can* be had *as whole* is the empirical surface. In this sense, even if one has "whole" access through a literal "play by play" recording of past events, what we have in total is still not the traumatic experience, the truth of the experience now muddled further by the presumption of thinking oneself as having access to the whole.

The idea that one might attain the whole can only be conceived by barriers erected in the conscious mind, the refusal to go into the murky symbology and narrative underpinnings involved with the unconscious. This suggests that insofar as a limited conscious keeps the unconscious at bay, it treats reality as an idol, what you see is packaged as what you get. If, on the other hand, there are layers to the "trauma" of beauty, then the unconscious proves vital in exploring those depths to garner what is true of beauty, in both negative and positive ways. It is in fact man's defensive reactions and etiological reactions to the difficulties surrounding the negative experience of

Beauty as trauma which have made the ecstasis and sustaining of its pleasure impossible.

The suggestion that consciousness formation, in its rupture from the unconscious, mimics the Fall of self from world, and of self from self, leaves only a half-stunned resuscitation from trauma as possible. Freud tells us that were we to access regions beyond the inert armor of consciousness, the steely inner layers would still shield against any excitation that promises a lasting effect[14] . Consciousness will flex its goal of receptivity by burning off each dying image before it is formed in memory as anything beyond the purely evanescent. Freud describes the consolidation of consciousness as characterized "by the peculiarity that in it…excitatory processes do not leave behind any permanent change in its elements but expire, as it were, in the phenomenon of becoming conscious."[15] As Freud states, "Becoming conscious and leaving behind a memory-trace are processes incompatible with each other within one and the same system."[16] Thus the symbiosis of consciousness and forgetfulness go hand in hand, and death is maintained as the most practiced and engraved right of creation; specifically the slaughter of all that is necessary to maintain consciousness. In the operation of integration, consciousness is built upon the dead.

The stimuli won't stop, and perhaps our fighting it is an attempt to control what won't stop, as if it comes from all angles as a torrent of endless new worlds to overwhelm the consciousness until beauty is the antinomy of peace, or a numbness that passes itself as peace. The act of the gaze, of aesthetic witness, which has a narcotic effect and is, according to Freud, a cathartic sublimation for misery in other parts of one's life, must always be secretly weighed against the destructiveness of its possible intensity and the construction of unbearable wish it might induce into the life of the imagination, as a derivation of sexual feeling.[17]

As Freud writes, in this approach consciousness is the avoidance of letting nature fissure through its defensive mechanism to that place of overwhelming pleasure that might manifest with the same force as sexual love. The nearly insensate layers are erected and parsed as an etiological reaction to the exploiting environment that won't conform or yield to the psyche, so it must save itself in a kind of reversal of unrefined, anesthetizing, mercantile economy as the frame of its happiness. Deflecting the beauty of the natural world to the extent that consciousness "serve to mask reality, rather than to reveal it."[18]

Thus, due to our constitution, there are endless worlds within such fragmentation, so that the new world is always new, purposefully ephemeral within the economy of suffering so that it may serve no more than evanescent pleasure. We are prone to not see beauty, but instead settle for pleasure and call that beauty. Its evanescence creates in us a longing of repetition by which the object of desire becomes more settled in its ontological echo as a thing, a belated thing, exploitable now for adapted yield of pleasure, rather than a relationship prone to the fragility of change.

The very notion of ecstasis suggests being able to see and endure beauty rather than become anesthetized to it. Consciousness is formed from a system of regulation incapable of ecstatsis, incapable of not becoming sedated to the presence of beauty under normal conditions, because such capacities have been deformed in the process of self-constitution.

Contemporary thinker Avivah Zornberg writes of an ecstasy in front of the idol that is a leaving of this world. Here we suggest that this taking leave is provoked by the ecstasy of the fetish. However unsuccessfully, the fetish tempts the seduced into total self-abandonment, but paradoxically, one feels a sense of control in this so-called abandonment because the object one has abandoned oneself to is simply that, an object. As an object there is nothing to fear, but there is also nothing to gain. Quite literally, we rouse ourselves into a frenzy over nothing, risking nothing and becoming nothing in the process, having abandoned ourselves to the nothingness of our own creation. Then we call this creation beautiful.

In describing the Pleasure Principle, Freud will allude to this relationship between happiness and beauty as incapable of lasting, despite the enduring beauty of the world. Due to the structures of consciousness itself and the way our organism is regulated, happiness can be no more than episodic. For if pleasure is prolonged, it does not culminate in stasis, but dissipates into mildness as if to protect itself against the encroaching force of inevitable suffering. "When any situation that is desired by the pleasure principle is prolonged, it only produces a feeling of mild contentment. We are so made that we can derive intense enjoyment only from a contrast and very little from a state of things. Thus our possibilities of pleasure are already constricted by our constitution."[19] Indeed, he goes on to agree with Goethe who warned that "nothing is harder to bear than a succession of fair days."[20]

Consciousness, deformed in its becoming by defending itself from the ordinary stimulus of beauty in its first consolidation, mars it with unknown trauma. Trauma is embedded in repetition and belatedness in the unwitting reenactment of the wound that cannot be either properly remembered or forgotten, but haunts.[21] This incapacity of consciousness to sustain pleasure in beauty due to a regulatory attempt to avoid (a forgotten) suffering points to a harrowing infidelity to serve the truth, even if seen. Further, trauma is forsworn to belatedness. As Cathy Caruth articulates, "This truth, in its delayed appearance and its belated address, cannot be linked only to what is known, but also to what remains unknown in our very actions and our language."[22] Trauma (as with *eros* and beauty for that matter) inscribes in us that what is *not* known is what we live.

It shows the degree to which we cannot contend that our consciousness will be able to rightly avail us of a necessarily probing investigation into the problem of climate change, as consciousness has not the wherewithal to think in the totalities that it ultimately abandons to politics or philosophy. Or rather, the totalities to which consciousness abandons itself are simply not totalities, ironically similar to the idol not being infinite, much less God. Beauty as we have conceived, and as a relation that is *not* finite, becomes too much to bear, thus we create the seemingly controllable fetish, be it of God or of nature. The paradox about the totality is that it is finite, otherwise it could not be a totality or "everything that is the case," to use Wittgenstein. Any attempt to grasp the whole in this way is an attempt to grasp the finite, the idol, and grasping *less* than the relation would, the relation that knows it relates to the infinite.

When unable to be reduced to use or pure utility, beauty is an open trauma as it draws us by *eros* into further exodus from ourselves towards that which can only emerge and be here a little while, its presence fragilely pointing to both what is and what is not, as the ecosystem of the infinite can only appear as fragment. As Camus articulates, "Beauty is unbearable, drives us to despair, offering us for the minute the glimpse of eternity that we should like to stretch over the whole of time."[23]

Constituted out of denial, consciousness is not equipped to transmit the fullness of experience and yet is always tempted to totalize experience. At times, and largely to its own peril, the psyche confuses repetition, especially within traumatic neurosis defined by its incapacity for genuine transformation, as an experience of wholeness, and proof of the exhaustion of the plenitude of possible experience. As Caruth articulates, "The experience of a trauma repeats itself, exactly and unremittingly, through the unknowing acts of the survivor and against his or her will."[24] Against our knowing and our will, we cannot see the world other than we want to because we are already suffering from the wound of beauty by which its originary experience imperiled us.

The self-deadening consciousness incurs to survive the primary unassimilable stimulation of beauty, deforms it, so that later its developed capacity to see beauty and attraction is accentuated around the sexual object, where efforts to fulfillment seem clearly delineated. Beauty, for Freud, is displacement, at best sublimation, of the sexual object.[25] The external stimuli of the world arrives in waves of such gigantic nakedness and by a force that would incur violence if we were not defended, and capable of deforming it into something that does not hurt us. We dismember the Giant, until the impulse of beauty becomes centralized, around its secondary sexual characters; its reduction to genitals is a vindictive act for its prior overwhelming of subjectivity. The Giant, the earth, though already largely dismembered as in the world of mythology, remains ever present as an inimical entity in traumatic neurosis, irreducible and essential, whose alterity is fantasized in unrealized and vindictive revenge for the prior self-sacrifice needed to dwell within it.

The sacrifice of the beautiful Giant, libidinal and quixotic, mimics the already absent self, a process that goes on without end, and is required to witness beauty that now can never really astonish again in the same manner because of our inert psychical armor.[26] In order for our capacity of sustained awe to be moderately renewed, we first need to change our relationship towards our own self-mutilation, whose wound only heals if it is kept open.

In this sense, Trauma is a memory-trace that comes in as a fragment to remind us of the apophatic, limitless whole, never born, but which sustains itself as the grounding orientation of our moral being, inhabited in its dependence, forgetting, and fragility, which can be re-invigorated to sustain us if we allow the fragmentation of trauma to continue to do the work of catharsis. This can occur when we allow and sustain the open wound whereby we once killed, and in killing keep killing the Giant, in the formation of ourselves as its opponent by way of the conscious mind.

The ambiguities and failure of consciousness to amend its traumatic errancy within aesthetic contemplation raises the following question: can beauty be reintegrated within us in a form that resembles profound desire, or is there yet in us a new world, a field where the work we do there can again resemble the fecund shadows of the "eternal pasture"?[27] Where regret and remorse fulfill themselves in a metanoia such that a turning away is a turning back towards the natural world?

At the same time, such repentance cannot be undertaken with the solipsistic aim of self-mastery, which unavoidably undermines the essence of a collaborative relationship between self and natural world. All life shares in biological material that erupts into traumatic imbuement. Perhaps there is a narrative space—and therefore a creative, imaginative space—of unclaimed experience, a new world where we are free enough of the manner by which we came into being, to see being. And in this realm of new and shared perception, we might again be pilgrims, exploring and tending a beauty that is not first shaped by our own death in a ceremony of grief and anonymity. Jorie Graham asks in her work *Erosion*: "How far is true enough? How far into the earth can vision go and still be love?"[28] Perhaps more sharply asked, how conscious can vision be and still be love?

If modern consciousness is the collaborative undertaking we have articulated above, aesthetic contemplation must proceed first from the unconscious. Consciousness is unable to see, or more acutely, envision, beyond the fixed symbolic order of communal desolation unless it listens to the trauma urgently articulated by ruptured fragments of the unconscious. This process of fragmentation supersedes the rational symbolic order, with its own logic of creation and vanishing, such that, as in Art, its inherent incompleteness can also be a source of catharsis. The unconscious proffers a way of seeing that is critical of seeing. The fragment liberates from the catalogue of the world, frees us from our fixation and habitude of a surface count, a kind of math by which capital trains us to see and dispart creation into sterile associations of "unredeemed realism."[29]

This habitude of surface is an attempt to unsee ourselves as fragmented and belated in our relation to beauty. The unstoppable rate of natural destruction in which we unwittingly participate can make one's very life feel unredeemable. Both *eros* and trauma entwine us in a webbing of belatedness, in which we are fragmented in a kind of double consciousness between life and death, caught by the feeling of it being too late, and with the object of desire ever receding out of reach. In terms of ecology, catharsis comes when we are willing to dwell and be our own belatedness, to endure its traumatic issue, which is experienced as an endless double telling between life and death, each appearing in their own fragmentation and rupture, where the urgency arises of not knowing if the trauma is the engagement with death, or with the ongoing experience of having survived it.[30] As Caruth writes, "At the core of these stories (,)…is a kind of double telling, the oscillation between a crisis of death and the correlative crisis of life: between the story of the unbearable nature of an event and the story of the unbearable nature of its survival."[31]

The inextricable story of survival and extinction, wherein the object of beauty continues to recede into both, is the story of our age. Catharsis comes when we perceive and acknowledge the wound and know that it cannot be separated out into separate possession, but links all to all. Any attempt to turn away from it only creates its further propagation and repetition. We must be willing to feel the wound, and not know rightly even whose to call it, or when it first occurred; as victims of our own trauma, even this experience will not be fully assimilated when it occurs, and we shall remain belated yet in our knowing of its true nature. This is now perhaps the beginning practice of *theoria*, or aesthetic contemplation: the seeing of the divine within all things. Inestimable joy comes from being in a participatory process where we again participate in a divinized world, and uncovering the Incarnation even within the mechanistic world of consciousness. As Dostoevsky's Elder Zosima says in *The Brothers Karamazov*:

> Love all creation, both the whole of it and every grain of sand. Love every leaf, every ray of God's light. Love animals, love plants, love each thing. If you love each thing, you will perceive the mystery of God in things. Once you have perceived it, you will begin tirelessly to see more and more of it every day. And you will come at last to love the whole world with an entire, universal love.[32]

Too often, we turn away from the infinity of the whole and substitute instead an idol to represent it. In the middle of Exodus, the chosen ones, engulfed by the feeling of belatedness, do not endure this feeling of it being too late, of feeling that both the earth and God have turned against their survival. When they feel the noose of the finite and the infinite closing in on them, they do not move in Exodus deeper into the wilderness towards liberation; they stop, they abort erotic catharsis and make an idol of their old longing for home. They deny the gift of belatedness where one does not attempt to separate, what Freud calls the crying wound, of traumatic neurosis, into its separate voices—instead one merely responds. In our times, picking one species to save over another, based on size or cuteness, is the making of an idol torn from its ecosystem. We must refuse such cheap catharsis and return wholly back to *theoria*, to seeing the divine revealed in each thing and how they interrelate even if fragmentarily. Our conscious ideas of keeping tallies on species by numbers, and only saving the chosen few, will lead to further decimation and corruption of the natural world, especially when it is clear that our scientific models have failed to keep pace with actual climate change, thereby making us further belated.

It is not simply that the fragment cannot be seen as a whole, but neither is it a shard excerpted from the whole in the manner of an idol parsed from context. It cannot be separated out as a smaller whole. The idol offers itself as substitute, as cheap catharsis, pulling itself up from the roots of creation to be immune to the transformation that eternity might still make us suffer. Such excerption, whereby the object of beauty is freed of its roots and environs, is not ultimately an exaltation of its uniqueness, but the advancing of its extinction. The people molded the Golden Calf "by excerpting a fragmentary reality—the ox—from the divine chariot empowered by four animals…the people select this image, with its field of memories, to represent the psychic intensity of the wilderness experience…by fixating on one image, they betray imagination…The Calf-ox is such an inhibiting image, confining imaginative power to a stable form."[33]

Artifact and idol belong to that species of realism that freezes the imagination by asserting a singular image in the urge to objectify. The idol is the projection of ecstatic singularity: the individual annealed, buoyantly immune to the materialism and erosion by which nothing remains its singular self on earth. A false catharsis of a solitude grows immune to the transformations of time; pointing solely to itself.[34] As something whose edges are nebulous, the fragment can at first seem little more than a ghost of its own dream, but the shard tells more than the story of its relationship with its own separation. The fragment, unlike the excerpted image, speaks for all. In its sharp incompleteness, it displays a doubleness, it mimics the longing of *eros* to project itself in front of itself to engage the other, to overcome the other as other.

The fragment exerts a kind of self-mastery, paradoxically, by vanishing. This does not mean withdrawing itself from the field of its origin, but rather pointing beyond itself, offering itself to that which it seeks to give freedom and expression.[35] The *techne* of the fragment works, like *eros*, with an intent to pour itself out without remainder towards the ever receding beloved, who cannot be in the name of love, possessed. The fragment resembles the terrain of the desiring soul. Simone Weil writes, "We unite ourselves to God in this way: we cannot approach him. Distance is the soul of the beautiful."[36]

The fragment works like an icon where one dwells in a space that does not feel like solid ground. The icon is meant to look flat, creating a paradox of the infinite wedged within such flatness that in rare moments of insight can reflect how everything springs from virgin dark to virgin light, from sleep to wakefulness, and then recedes again poised in the equilibrium of mystery, found in the face of the other. In such recession the viewer is called through the icon as the icon mediates immanence and transcendence, light and dark, past and future; so too in the face of the other.

Whereas the photograph and the simulacrum are like mortal wounds of enclosure based on the false notions of property and possession, the icon is a way of seeing that is critical of seeing, that refuses the easy catharsis of finitizing the things of the world, including the self. To finish the fragment would lead not to disclosure but to its deformation within a false finitude. Zornberg writes, "If the image that is present does not make us think of one that is absent, if an image does not determine an abundance—an explosion—of unusual images, then there is no imagination…and a stable and completely realized image clips the wings of the imagination."[37] When there is no imagination engaged by the viewer, there is only possession as intellectual object instead of relation. In this space, the way of idolatry begins.

If the image that is present does not make us think of one that is absent, it operates essentially as something extinct. The fragment incites the imagination to point beyond itself. It refuses the easy catharsis of being settled into definition of subject or object, or more importantly, of letting you know the details of its endangerment. It creates in the reader the longing to cherish and save what little is left; creating by its subtraction a negative ethics of patience where attention to the smallest detail of what has been saved, the remaining scrap somehow miraculously calved from the disappeared, is a form of prayer by which the I is also revealed in the beholding as the "not-me" that is worth saving. It takes time to develop, but love is the organ of patience. We wait on the one we are not born with but hope for, just as Yaweh provides a desert foretaste of what His later incarnation will feel like.

The icon is the wilderness. No path is straight. To wish for the straight path is to desire an idolatrous regression that only further enslaves. Once the pilgrims left Egypt, there was nothing to recognize. Night was beyond dark, and for a long while encrypted only with homesickness. The navigate stars above them held an authoritative urge to go impossibly back to what was gone to the other side of mourning. Eventually, in the dark, they learn to see the dark again. But by then it is very late in the story. In this way, belatedness marks cathartic ecology. We cannot go back, and we are short on time. The environment, like the wilderness for the pilgrims, was "born exactly at the moment it became the problem."[38]

We are belated. Worse, we are a fragment of belatedness. We must refuse the habit of the beautiful soul to graft the ghosts back on the limbs in nostalgic resignation, to really see where we are, to really see the wilderness will take longer still. Zornberg writes, "…a river can be represented in a photograph, which fixes its flow and makes it possible for it to be viewed and grasped. Yet the movement of the river, in its larger course, cannot be grasped in a moment. Rivers and selves, like music and narrative, take time to happen in."[39] Perhaps it might be said that the earth is a beheaded Giant that if we hold rightly, if we meet in the place of the thing, and not merely as an environment, where thing is a meeting and not a place, we might still see into God's vision.[40] Love, as the refusal to possess and is always a fragment, leads back into the plenitude of the imaginative powers. It's incompleteness, rather than failure, is the narrow trail to home, to the new world where we must find our dwelling. Home will not look like home at first. It will not be what we imagined. "We always think of the imagination as the faculty that *forms* images. On the contrary, it *deforms* what we perceive; it is, above all, the faculty that frees us from immediate images and *changes* them…Thanks to the *imaginary*, imagination is essentially *open* and *elusive*. More than any other power, it is what distinguishes the human psyche."[41]

It distinguishes itself as the sight of transcendence, wherein, and without need of complete narrative or exhaustive information, images swarm in their journey to fissure, dissolve, grope and adhesively reform themselves in their migration to leave the all too stable old world for the new world whose vision is bisociative and more like God's, who at one point in midrash interpretation confesses—"You see only one sight, but I see two."[42] Like the angels who are said to swarm lovingly around the host with wings of a thousand eyes, the deeper one pushes into reality, the more manifold the kinds of vision and narrative insight become.

Whereas the idol is the product of fidelity to a single and stable cognitive infrastructure, the alternative, bisociative vision, involves a doubling process of inwardness and outwardness, wherein suffering and pleasure are once again bestowed to the things of the world, as oneself is bestowed to it. We must let *eros* split us further, a fragmentation of the single narrative so that the plenitude can once again cross over the wall of sterility, remoteness and human desolation. "The alternative, God-taught model for vision involves a doubleness, a desire for otherness. The secret of the One cannot be seen except where there are two."[43] In practice, it is a doubleness of sight, a singularity of intention. Such vision seeks communion, and sees earthly inhabitants as conjoined inheritors born from a billion years of stardust, appearing in such plenary multiplicity that is the result of God's mysterious dream. When we step back to see the common world born from the scraps of stars, it is impossible to think of ourselves so different than other creatures, or to want to continue to break communion through objective categories that ultimately induce violence.

The doubleness eventually points to what will end as erotic failure, as incapacity for full union, where *eros* cleaves back again into its disparate components of pain and pleasure. Eventually desire renews itself and takes on the mourning shape of what it wants and lacks. *Eros* is marked by lack; compassed by dearth and insufficiency.[44] We must love it enough to know we don't know. We must love what we don't know. We must let it have its story even when it incriminates us. We must be erotic forgetters to be erotic seers. Of God's vision we need not even be religious believers to begin to value the complexity of the natural world differently. For instance, Morton writes,

> "Instead of serving up lashings of guilt and redemption, might ecological criticism not engage the ideological forms of the environment, from capitalist imagery to the very ecocriticism that opposes capitalism? Ecocritique would establish collective forms of identity that include other species and their worlds, real and possible. It would subvert fixating images of 'world' that inhibit humans from grasping their places in already historical nature. Subverting fixation is the radical goal of the romantic wish to explore the shadow lands."[45]

Love, as a fragmentary relation to what is and is not, leads us back into the plenitude of the imaginative powers to engage the unknown. At times it is a residue of a fire that we feel like a timeless heat, a fever in our bones. A new world suddenly ruptured within the old one where catharsis comes only from sustaining oneself within the fire where desire suffocates, where it stirs up a lack of being, recalling the trace of something more. There is a story in the midrash tradition where apophasis is felt physically as kind of illness, a fever in Moses's body, which is later on called prayer. He carries in the fire of his bones, the desire to save creation when he thinks God wants to destroy his people. It is as if he carries the letters that were written on the first broken tablets, written like a kind of fire into his physicality whereby God reads back his own greater, but at the moment, subconscious desire to preserve the world. As is written in Jeremiah, "I will put my law within them, and write it on their hearts."[46]

"Now leave me alone," God says to Moses when he sees the infidelity of his people, "so that my anger may burn against them and that I may destroy them. Then I will make you into a great nation." When God threatens to destroy his people, in the midrash accounts, Moses pleads for God to kill him instead of saving him as the start of some new chosen race.[47] As a form of entreaty, Moses comes down with fever; rather it is said even more poignantly in the Zohar's midrash account that he is plagued with a fire in his bones; and in another account he becomes "ill with prayer."[48] His unvoiced protest is his illness, it settles in him like a smoke writing on the tablet of his bones to entreat God of his earlier promise to save his people, which predates their invention of the idol. "The intimate human dream of fire," as Zornberg writes, "is in the language of dream that the midrash speaks when it evokes the desire for union and the fear of being consumed."[49] Moses embodies the apophasis of that unseeable union wherein he risks total consumption in an attempt to attenuate the extinction of others. His catharsis is his illness by which he achieves a self-forgetting, and literally utters from the bones of his skeleton, where deepest human identity holds. The fever is the prayer of seeing and saving. The illness provides the self-forgetting necessary to dwell unconsciously, with his people even when separate from them.

This provides a path to heal and recover from the projection of God as a God of vengeance, literally worked through in the dark of Moses's body. Through a process of pathological mesmerization, Moses's body briefly becomes the promised land wherein he and God meet and project the continuation of His people. That is to say, it inscripts itself into the old definition of thing as meeting place wherein he is willing to sacrifice himself for the succor of others. His pathology represents his people's pathology. In the old work of being a scapegoat, he inhabits their sin before the Beloved and they meet in the field of the future. They meet where the world is burning in the fire of his bones, as if it is the great wasteland where he and God go to see the extinction of his people. There, he and God can imagine what it looks like, they can project and rescind his perception of God's desire for extinction, and, as in the case of Adam, remake a people from the bones in order to ensure a re-emergence of meaning and value.

The fever brings on a self-forgetting, a kind of possession that enables him to void the temptation of idolatry, where he and God can be erotic seers, where the promised land seeps into the very heart of the gift of the illness that will not let him settle. It offers an unusual glimpse at beauty that reaches far below the surface. Simone Weil writes:

> …we want to get behind (the) beauty, but it is only a surface. It is like a mirror that reflects to us our own desire for good. It is a sphinx, an enigma, a sorrowfully irritating mystery. We want to feed on it, but it is only an object we can look on; it appears to us from a certain distance. The great sorrow of human life is knowing that to look and to eat are two different operations. Only on the other side of heaven, where God lives, are they one and the same operation.[50]

Within Moses, this operation occurs by a foretaste of fever wherein he and God project into the future; that is to say they arrive belatedly to the field of His dreams and see only the Dead. Out of love, Moses burns with a presentiment fever. His desire to save his people sits inside him like a war. As Anne Carson describes *eros,* "The act of loving is a mingling and desire melts the limbs. Boundaries of body, categories of thought, are confounded.

The god who melts limbs proceeds to break the lover (*damnatas*) as would a foe on the epic battlefield."[51] God and Moses look at the extinct. The horror of what is seen is broken down into an insensible digestion within Moses who must metabolize the war that his conscious mind believes God desires. His bones must burn with the undigestible horror of God's perceived wish until Moses experiences a turn away from this wish and back to God's original desires. It is as if God uses the very bones of Moses to speak: "I can say that I couldn't feel the slightest difference between this dead body and mine. All I could find between the dead body and mine were obvious similarities, do you understand?"[52] The rib that God jangled from Adam to become Eve, in one of the two creation stories, is the foremost underlying human identity. It recovers the human project, otherwise hidden and unspeakable by any other language to God. The bones engulf both the prophet's objectivity and his subjectivity until God is filled again with desire when he imagines His people's extinction. "Whoever desires what is not gone? No one. The Greeks were clear on this. They invented *eros* to express it."[53] In the fire of the prophets' body, in his absolute solitude, the world dwells again in truth. The imagined extinction brings Moses back to God's original desire.

Along these lines, it is the imagination which allows past trauma to have meaningful voice "in our bones," as well as force us to feel the dread of possible extinction, overcoming denial and reorienting us to nature more authentically. We are in this sense in and out of time, historical and ahistorical at once, in that our ability to be fully present rests simultaneously in our capacity to let go of the present, at least enough that the imagination can play its part with past and future. But it is critical that in allowing the imagination its role, we have to also allow that this is sometimes experienced as pathological, as our conscious mind fights with us to keep out what the imagination aims to let in and mediate between, the unconscious providing past trauma and well-founded future fears that the conscious mind cannot fathom.

In the end, God recants and Moses descends back down the mountain to his people. As the story goes, Moses does not reach the promised land. And while this is presented in scripture as punishment, it is also his salvation. Joshua reaches the promised land and abides the killing of women and children as justified for possession. Moses only sees the promised land from a distance, is allowed to admire it, contemplate its riches and beauty, but never possess. Violence makes no sense when possession is not at stake, but rather contemplation. More than this, in contemplation he does not presume the whole; he knows he only sees in part. This knowledge gives him more of the relation than the presumption of the whole gives the one who conquers in possession, no longer able to contemplate beauty, but merely own. Moreover, what is owned is dead.

Each of us, in the linking back to Moses, comes to the wasteland where love hurts, where we see what is gone, disappearing into the annals of the dead, where every nine minutes a new species falls beneath the century. Our belated perception of a gift now irretrievably lost erupts in grief, and we are fragmented and sickened by what we see. We grow weary of the body, wearier still of being human, with our inheritance of violence, that we try and flee the wasteland by aiming instead for self-mastery that attempts to once again control God's vision, rather than be in God's vision.

We do not need to master ourselves as a means of evading despair or conflict; mastery is not the pre-requisite for re-visioning history. In fact, such a quest, as we will see in the next section, more often than not, leads to further devastation.

APOPHATIC CATHARSIS

The catharsis of the fragment emerges because it points to a wholeness that eludes reason and remains unattainable. Smoking in the bones was his prayer, was his fragment offering of an unspoken conversation with God regarding the sustained continuation of Creation. Where his prayer, his fever, instantiates a new world that is neither his nor God's vision.[54] The bones burning inward with a stigmata of prayer, not unlike an icon in that the icon is a way of seeing unconsciously, neither property, nor possession.

Moses descends down the mountain to find his people praying to the idol. Unlike his transformative fever, "They dance in timeless ecstasy in relation to which reflection and re-creation are foreign modalities."[55] The ecstasy in which Moses finds the children of Israel before the calf is not one of true self-abandonment but rather a solipsistic absorption, in that they abandon themselves to an object. In this sense they abandon themselves to themselves, a sort of perverse self-worship, owing nothing to this statue that they themselves constructed. They celebrate that of their own making, a veritable tower of Babel, erected this time in the woods.

One can imagine that without Moses, they may have gone further still, erecting all forms of beauty without nature, against nature, no longer wanting that apophatic relation to nature. They might have gone on to put their names on their new buildings, erecting idols to themselves in pure solipsism, no longer on a journey into the unknown and unknowable.

A similarity between such self-absorbed ecstasy and rigid piety is this: they each wish to petrify the object of worship. They are only seeing, not allowing themselves *to be seen* by anything. This inability to be seen is indicative of the lack of vulnerability in each approach. One is like a drug-induced stupor, the other a controlled and deadly sobriety, the latter fearing any change of consciousness whatsoever. But in both cases there is no abandonment because there is nothing to whom to abandon oneself and thus nothing to receive. One must get out of one's own way to receive.

When abandonment does occur, deep catharsis comes as an opening rather than a closing off of the world, it appears as an Exodus from consciousness and, as such, it necessarily presents itself as the crooked path. This crooked path is the path of Love. God leads the people in a crooked way through the wilderness, taking over forty years, so they could learn to see again. Patience is an abiding of love's crooked ways. We need an exodus from self, from our conscious identity and perceptual habits to again open up and see the habitat.

For Moses, the fire of the bones is like an internal icon, the anxiety we carry with us, that based on the most intimate desire of union and fear of consumption by the flame. Zornberg writes, "We are speaking of the intimate dream of fire. It is in the language of dream that the midrash speaks when it evokes the desire for union, and the fear of being consumed.[56] This fear of being consumed is our most intimate relationship with beauty, the reason we first hardened ourselves against it, now we carry the fire of consumption already in us. The inwardness and the outwardness of this erotic transit has the same relationship as the icon to the seer. Moses is seen from inside the deepest place in him, a fire is started, as if the rest of his muscles, sinews, organs become the projected land through which he and God walk so there is no dissimilarity between inside and outside; all has become the holy ground of the Burning Bush. The pathology allows Grace where the fire seeps together the particulars so that each thing is seen, as Rodin writes, "as the limit of the flame to which it owes its existence."[57] This is to say each thing is seen iconically, in which now the whole world is burning with God's uncreated fire visible only to those who, in humility and belief, witness the earth as that which is holy.

NEW WORLD AND AESTHETICS

Of the beguiling moments of encountering the New World, F. Scott Fitzgerald wrote, "Man must have held his breath in the presence of this continent compelled into an aesthetic contemplation that he neither understood nor desired."[58] What they saw before them was so wondrous and unexpected that it hurt, exceeding any and all possible expectations. This moves against the scientific theory of the time, which had advanced that the animals of the New World were vitiated and degenerate copies of those of the Old World.[59] Eventually, Thomas Jefferson would send Meriwether Lewis and William Clark to find the Northwest passage, with specific instructions to describe, if not procure, the bones of the wild incognita, that he imagined still "remain in their aboriginal state, unexplored and undisturbed."[60] For some, like Audubon, the early imperial incursion into the new world revealed a choked plenitude that was at once awe-inspiring and brought with it tremendous anguish. He writes repeatedly in his journals of the pointless mass slaughtering of heretofore known and unknown animals by his fellow frontiersmen; men who, in absurd fashion, would leave the slaughtered remains to smolder in a burning world on the banks of the Missouri as they oared away.[61] Watching the prolific waste of animal blood secreting into the waters as the very antinomy of Theophany must have inspired tremendous distress in a man who would privately pen, "A true conservationist is a man who knows that the world is not given by his fathers, but borrowed from his children."[62]

Later, Audubon would fulfill a stated need to "still" bird specimens: to kill the animals, to position their bodies into a recalled animation and to copy their likeness. The enclosure of the new world into uninhabitable book is not without its own imperialistic tendency for possession at any cost. This need to kill to create the substitution, if not simulacrum, shows that even aesthetic contemplation is not always benign. Even though he didn't want to kill, in order to achieve a kind of mastery wherein the dead specimens were to appear life-like is inseparable from a kind of self-mastery he pursued in his art. "To have been torn from the study would have been as death; my time was entirely occupied with art."[63] Later he articulates, "Birds have been the gadgets of my maximum delight."[64] It seems that violence in the service of mastery was somehow permissible.

The birds remained objects. The birds, as gadgets, became a kind of manifest destiny for him and the means by which he created a kind of self-mastery through aesthetics; when you lose the fact that you see as you are seen, even as you try to master the art of representation, there is an imperialist component that is as dreadful as any other kind of manifest destiny. Audubon writes, "But the moment a bird was dead, no matter how beautiful it had been in life, the pleasure of possession became blunted in me."[65] And yet, he did kill, and killing served his "art" such that his love of nature was cleaved from his capacity to represent nature, and he chose the act of representation over any other kind of *eros* that he may have shared with nature. In the end what he produced lacked that iconic relationship with nature because it became atomistic in his work, not unlike the excerpted golden calf. Zornberg writes, "By fixating on one image, they betray imagination…God reproaches Moses for foreshadowing the human need for bedrock, for the stable form of a single cognitive frame."[66]

The rigidness of Audubon's birds points to a paradox of the imagination where such rigidness appears to be a deformity of the image precisely because of its attempt at exactness wherein such exactness is lifeless precisely because the imagination did not try and transcend the image before it to make appear real and animate.[67] In order to create a true link to reality, the imagination works best when it *deforms* what we see. Correct deformation, which might even be considered as a kind of forsaking, is actually the re-forming of the image through a fragmentation whereby the image transcends itself towards the animate by pointing to its incompleteness. Bachelard sites the imagination as the space wherein images are deformed rather than formed, going so far as to say that if the image is not changed, but merely objectively represented, there is no imaginative act. Rather, the imagination is the faculty of novelty, it links us, like the unconscious, to what is essentially evasive and uncatchable. Bachelard suggests, "It is the human psyche's experience of openness and novelty. More than any other power, it is what distinguishes the human psyche."[68]

Against the technical difficulties of both memory and perception, which produce a feigned stability, imagination reanimates the life by a process of deformation. It is what Avivah Zornberg calls the questing imagination, wherein, "Images, in all their presence and power, are constantly transcended by imagination…to respond to the rigors of a questing imagination; it is the faculty of deforming the images offered by perception. It involves swarms, explosions, unexpected union of images; but it pulls loose from the "all too stable background of our familiar memories."[69] In other words, the imagination, when it transcends the image, liberates us and the object of perception, into a life that is again shared and elusive, resonate most especially in the epiphanies of the beautiful.

One can imagine other ways to produce Audubon's work that may not have entailed such copious death, methods wherein he did not have to submit his love of nature to his need of perfection.[70] Aesthetics is not free of an imperialist brutality to fragment and betray the beauty it finds. It can be a cruel ethic, capable of seducing itself, under the guise of self-mastery, such that its means and ways are not necessarily other than mercantile.[71] But aesthetics needs something to mediate its execution, otherwise our attempts at mastery are going to end up being life and earth denying, and not unlike manifest destiny, wherein it is believed a God-given right to conquer the Giant of the land at all costs. Aesthetics *as the result of mastery*, particularly the ideology of representation that presupposes itself to be less violent, denies its origins in trauma.[72] Repeating through the act of killing and then representing the bird as if alive and flying, as in the case of Audobon, the originary framework of consciousness may itself be said to be born out of death.[73] In *Unclaimed Experience*, Caruth argues that in Freud's later theories, consciousness is married to death from its inception, and that Freud moved from seeing trauma as episodic or accidental to being foundational to the very awakening of self. Caruth sums his later theory as "an explanation of the origins of life itself as an 'awakening' from death that precisely establishes the foundation of the drive and of consciousness alike."[74]

If consciousness is born of enwombed death, then we arrive as belated things into the real, wherein the aesthetic goal of self-mastery is to not be a living spectacle of death. Consciousness denies its origins, unable to recover how its origin and its secondary defensiveness against its origin affect its drives, intentions and perceptions. Instead, it tells itself a story of coherency, an ideology built on the illusion of progression from animate immaturity to animate maturity. Its lie: an evolution in which we are born to die, rather than die to be born, is meant to create the illusion of conquest; of ascent, rather than circular regression back into the place from which we came— to die once again inside the encaved body of a Giant whose death, if secured, might mean our freedom.

A repressed and unvoiced aggression against the earth therefore is bound in the concept of self-mastery, best depicted in the mirror stage.[75] Sunk in helplessness and nursling attachment, subjectivity nevertheless creates an ideal, an imago, independent of the mother/world holding it up. The ideal will become paradoxically rigid, reminiscent of the inanimate shell from which it originated in first consciousness, to tell the story of spurious self-mastery, wherein even the past is reabsorbed, retroactively imagined within the de-forming faculty of the imagination so that consciousness originates from being, instead of the gravity of death."[76]

Inevitably, however, the trauma of origin re-emerges. The birds, in this case, remain gadgets of Audubon's self-mastery in a repetition trauma of bringing something out of death. Even in aesthetics, such repetition compulsion brings further violence, if not extinction, upon the objects of desire as we try to secure our grasp within the world. Zornberg writes, "Self-mastery is, in some unavoidable sense, the realization of the falsity of the self… This moment, then, is a 'brief moment of doomed glory,' like that of Adam and Eve, expelled from paradise into history: anticipating mastery, they actually gain a 'horrified recognition of their nakedness.'"[77]

Belatedness is our truest nakedness. Relation, rather than mastery, must be established out of this impossibility of how late, and how unknowing of it, we arrive. Were consciousness an original awakening from the death of beauty, we might be able to respond in time. Even if not, the gift of belatedness would tell us to respond clearly and consciously, no matter what, to whatever cry we might hear within the enigma of trauma. However, to the extent that consciousness obscures reality more than disclosing it, Caruth's assessment rings true that the wish fulfillment of consciousness is to not wake, to abide asleep, and to go on sleeping."[78] As she states, "It is not the father alone who dreams to avoid his child's death, but consciousness itself that, in its sleep, is tied to a death from which it turns away."[79] To meet the world, as Jorie Graham does in "Sundown": "I am not like a blind person/ walking towards the lowering sun,/ the water loud to my right,/ but like a seeing person/ with her eyes shut/ putting her feet down/ one at a time/ on the earth."[80] We do not see because we sleep, and in our sleep, we continue to think we are on time. We need the dream to wake us. This brings us again to the place of cathartic ecology.

In the wilderness, catharsis came first for Moses as sickness, then as epiphany of not knowing where his unconscious ended and God's began in the fever of wanting to save creation. The fever becomes a holy place that marks him internally to remind him that wherever he walks, as before the burning bush, is holy ground. The bones became the tablet of a physical and intimate dream between creature and creator, where the icon is born as the artifact of the unknown, remaining unknown as it is known. The icon wakes us, but it wakes us in the unconscious. Here it is humanly possible to be on time. In the unconscious connection between ourselves and God, man is no longer a living spectacle of death, nor is the inanimate our most intimate umbilical cord; rather, he is the intimate fire necessary for creation. The catharsis of fever is the iconic language of intimacy where Moses continues to live inside the dream as it is written onto his bones. His illness transitions into rapture, where it operates in its writing as a means of being seen and seeing, as the dream deepens into God's original vision, where the plenary is loved. In dream, Moses first achieves exodus; his perceptual habits transformed to again see the habitat. By the fever in his bones Moses learns firsthand that his own embodiment is erotically entwined with the divine. He succumbs to a revelation that is literally felt in his marrow, and he awakens later knowing that above all else, the Image he carries is foremost intimate and goes beyond the merely ethical while still including the ethical. Knowing that he cradles in his bones traces of the invisible makes it possible for Moses to see the visage of God in the visage of earth itself so that his connection to the wilderness is born, at the very least, of the ethical primacy of Thou, and not based in authoritative or epistemological desire and transcends the ontological demise inherent within both.[81]

THE BURNING CHILD

In *The Interpretation of Dreams*, Freud writes famously of the dream of the burning child to explain the latency of traumatic grief. When trauma is severe enough, subjectivity masks reality; in this case, the death of a child, so that the father must go to sleep to see reality in which the child speaks to the father about his death. The living presence of the dead child talking about his death will beset paradoxically upon the father as the desire not to wake up.[82]

Freud writes:

> The child's body was laid out, with tall candles standing round it. An old man had been engaged to watch over it, and sat beside the body murmuring prayers. After a few hours' sleep, the father had a dream that his child was standing beside his bed, caught him by the arm and whispered to him reproachfully: 'Father, don't you see that I'm burning?' He woke up, noticed a bright glare of light from the next room, hurried into it and found that the old watchman had dropped off to sleep and that the wrappings and one of the arms of his beloved

> child's dead body had been burned by a lighted candle that had fallen on them.[83]

Freud goes on to adumbrate the wish fulfillment of the dream, saying:

> The dead child behaved in the dream like a living one: He himself warned his father, came to his bed, and caught him by the arm. The dream was preferred to a waking reflection because it was able to show the child as once more alive. If the father had woken up first and then made the inference that led him to go into the next room, he would, as it were, have shortened his child's life by that moment of time.[84]

So it is in dream that conservation occurs. Even where it is an illusion, the dream speaks to the wish that the child not be dead. It outlines both the wish and the truth that consciousness can only come to us belatedly. As Caruth articulates, "to the extent that the father is awakened by the dream itself, his awakening to death is not a simple matter of knowledge and perception but rather, Lacan seems to suggest, a paradoxical attempt to respond, in awakening, to a call that can only be heard within sleep."[85] Thus, the awakening is an exodus from consciousness where the wish fulfillment remains to stay asleep. The terrible brutality of the child's death makes the father late to reality. Such belatedness characterizes exodus, where the unconscious has to mediate the truth that consciousness begs to escape[86].

If we replace the burning child with the burning world, this call to awaken that the conscious mind does not want to hear is because it pursues its wish fulfillment in sleep and takes on a new dimension about our capacity to respond. As Frank Seeburger writes, "Such awakening can only occur belatedly, only 'after the fact,' namely, after the call to awaken has itself died away into silence insofar as it cannot be heard—cannot be heard precisely because there was no one awake to hear it, any possible hearer having fallen soundly asleep, and thus necessitating being called to awaken."[87] But it is still better than violently remaining in the conscious. This is the more hopeful path, the best chance we have, to carry out the imperative of repenting from a collective solipsism. "Kiss the earth, flood it with tears, ask forgiveness!"[88] The asking of forgiveness of what does not speak a shared language can only be communicated as the direct action of love. Such repentance is born with the urgency of sacrament, rather than mere commitment, that nothing more go ahead of us into the rooms of replete dark and injustice.

It is fair to say that what we don't wake to in time still goes on defining us, still goes on trying to wake us with its own fading cry. The depth of extinction, in which a species continues to disappear every nine minutes to the darkness of history, is a nightmare we don't want to wake to— where awakening from sleep is the same as an exodus from consciousness into a totality so grievous that it feels simultaneously that there is nowhere to safely awaken and no one to safely wake. But this is the realm of cathartic ecology, the realm that cannot be heard "precisely because there was no one awake to hear it, any possible hearer have fallen soundly asleep."[89] Such belatedness awakens us into the violent fade of echoes, where we are but fragments of who we feel we need to be, to adequately respond.[90] "Thus awakening from sleep—which simultaneously means awakening from consciousness, is itself traumatic. One falls into consciousness as into sleep, and can only be awakened after the fact by a call to awake that can never, in principle, be heard when it calls, but can only be heard in the recall, as it were, of its echoes."[91]

Whereas consciousness is obsessed with totality, the unconscious is obsessed with unity and its placement within nothingness. Those who awaken are free to be and see fragments of a lost subjectivity. Further, it conserves and links father to son through the firelight of the candle that burns now also in the father's heart where the dead child is its most prominent ember. The unconscious can present and be a fragment of the real by which one can fall further into the abyss of the whole; that is to say, further into God's vision. Catharsis is what might be bequeathed to the survivors of subjectivity where "who remains is whoever/whatever comes then, 'after the subject,' to use a 'postmodern' formula."[92]

Being lost defined an entire generation in the wilderness who eventually learned to recover their sight. So while beauty is traumatic for consciousness, it remains a mediating form between God and man unconsciously to remind each, through iconographic sight, that the Good is beautiful; Creation should continue moving deeper into its unfinished icon and not be destroyed. Moses carries briefly for God the fever of desire for his creation that cannot be extinguished at a time when God wants to destroy his Creation.

The kind of mastery that we need in this moment is one that frees itself from the idea of awakening, of being driven towards the empirical tendencies of totality that have been argued elsewhere as ideology.[93] We need the sleeping subject to see the potency of the squandering task to awaken on time; our awakening is always already too late, but we must do it belatedly anyways. Indeed, we are belated to the burning both outside us and inside us. Failure likely surrounds us, as in Lacan's interpretation of the father waking inside the dream of the burning child. As summed by Caruth, "To awaken is thus precisely to awaken only to one's repetition of a previous failure to see in time."[94] But the words spoken inside the dream create significance beyond chance, where even though the possibility of response is again met with failure, it creates an ethic of continued response to the vulnerability of beauty, one in which the life of the survivor, who survived his own exodus from consciousness, is linked to the death he finally witnesses, and which frees him to act with an urgency born of living iconically from the death we already carry. To live iconically by means exceeding ethics in its plenitude of desire and loyalty by which we feel in our bones the irreparable loss of each extinguished species. As Foltz rightly notes, it is not about the aggregate loss of beauty or the number of specimens lost for study, but is the walling of a window in which the invisible manifests: "It is, rather, the progressive walling of windows, windows that we did not build and cannot replicate, and that will be walled over forever."[95]

Belatedly, in the dream, the father will wake and again fail and repeat the failure that he would rather die than repeat. Extinction carries with it an echo that goes far beyond what silence can compass. As in the transmission of the burning child, these are not facts that can be mastered. The easy death of what took millions of years to arrive is nothing short of a nightmare. Our tendency towards belatedness points to the tragic possibility of encountering the endangered as endangered because they are already extinct by the time we arrive, unless we learn to dream ahead of our own failure. We are located in the paradox that consciousness is asleep, but we have no other recourse than to sleep to try and see as ones who cannot see, because we sleep.[96]

It is in dreams that we escape the ideology of the surface and meet God again in the burning fire. As in Exodus, patience and eros will be required in what looks undoubtedly to be a pilgrimage into tragedy. Moses carries this sickness as a gift from God, it seems, and if he does not ignore it, he will do what is best for his people, and what God wants for the people.

No dream can be reduced to its facts, it is only in the transmission from that naked and burning other, that he sees himself as one who sees. Beauty is both the trauma and the saving grace, if we can withstand it. Its infinity can only be experienced and known by entering the unconscious. This suggests that beauty is often not pleasurable. It is a wound that we have to let ourselves feel and not attempt to control. The empirical approach is analogous to the conscious approach, as it presumes access to the whole, and even when not claiming to presume the whole, it shall remain problematic inasmuch as it presumes that consciousness is not impacted by the unconscious, so there is no need to include it in empirical "consciousness" studies either.

One becomes, in such awakening, free of consciousness. However, this undertaking requires an extended exodus from consciousness, and it is not a straight path, and in its winding mimics love. The only help is an exodus from self, from our conscious identity and perceptual habits, to again see the habitat. For now, the crying wound of trauma can be heard most clearly in the unconscious. We need to feel that the figurative fire in the bones of knowing love, just like in the love of beauty, is a fragmentary relation of what is and what is not. Beauty is a trauma; beauty is a saving grace. It works as an iconic sight to remind us of the goodness of creation, whether it is through pain, sickness or pleasure. This is the sight of catharsis.

CONCLUSION

We must stop using the conscious mind as a means of armor against the wound of beauty, and let ourselves be wounded; part of that wounding is an act of being seen. Once we feel ourselves to be seen as something deeper than mere identity, but a fragment of the whole—a process which happens unconsciously as the most heightened sense—then we are less apt to tear ourselves and the earth apart.

Most of our love for nature is unconscious, and one should not aim for self-mastery, whereby it is presumed one has a grip on what is seen, for that would be to treat nature as an idol. The icon keeps us from trying to enslave the beautiful, and allows for the catharsis of letting nature be; we acknowledge each portion has its own narrative in the story of the earth; to snuff out any piece of the story is also to crush the icon.

Once we feel ourselves to be seen as something deeper than mere identity, but a fragment of the whole, it is possible to realize that you witnessed wrongly the world that found you. Such realization may arrive as grievous, but the elegance of grief is that it can be momentous in its givenness, and monstrous enough to change you. Such woe can alter the assumptions by which you have formed your reality in a society that reassures you no massacre is committed in your name; every act and form of violence is permissible in the name of humanity which has so far produced nothing less than the death of the heart. If we come to terms with our existential belatedness, we can still, from the mysterious depth of being where love resides, engage in comprehensive and powerful direct action, wherein we protect that which we are grateful for, because the assumptions of our reality have been changed by our clearer vision.

Because Moses was able to carry the trauma of beauty and retain it in relationship to God, it might be said that Moses was the only one to reach the Promised Land because he related to it iconically in contemplative distance and not possession. This was a relationship in which he was willing to risk the peril tacit in the other's presence, be it person or land, and lead his people deep into the mystery of the wilderness by creating a means of liberation from the old ways, and without precision of destination. He reaches the Promised Land first and foremost in his imagination, learning the ethic of care for all creation through the willingness to incur suffering within his own flesh, an inseparable part of the landscape, and without any possible substitution.

BIBLIOGRAPHY

Audubon, John James. *The Missouri River Journals of John James Audubon*. Lincoln: University of Nebraska Press, 2016.

Maria R. Audubon, with Zoological and Other Notes by Elliott Coues, "The Labrador Journal." In *Audubon and His Journals Volume One*. New York: Charles Scribner's Sons, 1897.
Barzilai, Shuli. *Lacan and the Matters of Origin. Stanford:* Stanford University Press, 1999.
Bonhoeffer, Dietrich. *The Cost of Discipleship.* New York: Simon & Schuster, 1995.
Camus, Albert. *Carnets: 1935–1942.* New York: Hamish Hamilton Ltd. and Alfred A. Knopf, Inc., 1963.
Carson, Anne. *Eros the Bittersweet: An Essay*. New Jersey: Princeton University Press, 1986.
Caruth, Cathy. *Unclaimed Experience: Trauma, Narrative and History.* Baltimore: Johns Hopkins University Press, 2016.
Dostoevsky, Fyodor. *The Brothe rs Karam a zov*. New York: Dover, 2005.
———. *Demons*, translation by Richard Pevear and Larissa Volokhonsky. New York: Vintage Classics, 1994.
Duncan, Robert. "Often I am Permitted to Return to a Meadow." In *The Opening of the Field*, 7. New York: New Directions, 1960.
Fitzgerald, F. Scott. *The Great Gatsby.* New York: Simon and Schuster, 2003.
Foltz, Bruce V. *The Noetics of Nature: Environmental Philosophy and the Holy Beauty of the Visible.* New York: Fordham University Press, 2014.
Freud, Sigmund. *Beyond the Pleasure Principle*. New York: Dover, 2015.
———. *Civilization and Its Discontents.* New York: W.W. Norton and Company, 1961.
———. *The Interpretation of Dreams.* New York: Macmillan Press, 1913.
Graham, Jorie. "The Age of Reason." In Erosion, 19. New Jersey: Princeton University Press, 1983.
———. *Place.* New York: HarperCollins, 2012.
Hinton, David. *Existence: A Story*. Boulder: Shambhala, 2016.
Jefferson, Thomas. *Notes on the State of Virginia.* Boston: Lilly and Wait, 1832.
Kolbert, Elizabeth. *The Sixth Extinction: An Unnatural History.* New York: Henry Holt and Company, 2014.
McGilchrist, Megan. *The Western Landscape in Cormac McCarthy and Wallace Stegner*. New York: Routledge Press, 2010.
Mead, Rev. D., ed. "John James Audubon." In *The American Literary Emporium, Or Friendship's Gift*. New York: C.H. Camp, 1848.
Mitchell, Stephen A. *Hope and Dread in Psychoanalysis.* New York: Basic Books, 1993.
Morton, Timothy. *Ecology Without Nature: Rethinking Environmental Aesthetics.* Cambridge: Harvard University Press, 2007.
Plain, Nancy. *This Strange Wilderness: The Life and Art of John James Audubon.* Lincoln: University of Nebraska Press, 2015.
Sadosky, Leonard, ed. *Old World, New World: America and Europe in the Age of Jefferson.* Charlottesville: University of Virginia Press, 2010.
Seeburger, Frank. *The Open Wound: Trauma, Community and Identity.* CreateSpace Independent Publishing Platform, 2012.
Sherrard, Philip. "Human Image, World Image: The Renewal of Sacred Cosmology." In *Towards an Ecology of Transfiguration*, edited by John Chryssavgis and Bruce V. Foltz, 211–12, 221. New York: Fordham University Press, 2013.
Vedder, Lee A. *John James Audubon and the Birds of America: A Visionary Achievement in Ornithological Illustration.* San Marino: Henry E. Huntington Library and Art Gallery, 2006.
Weil, Simone. *Gravity and Grace*. New York: Routledge Classics, 2002.
———. *Waiting on God.* New York: Routledge Revivals, 2009.
Zizek, Slavoj. *The Sublime of Ideology.* New York: Verso, 1989.
Zornberg, Avivah Gottlieb. *The Particulars of Rapture: Reflections on Exodos*. New York: Random House, Inc., 2002.

NOTES

1. The history of climate change denial is extensive, and if one includes overpopulation as a main factor, biologists have been warning us of this for centuries. Sadly, the history of climate denial will likely become its own branch of study within either Trauma studies, Ecology or History. Such trenchant obfuscation and denial has been most recently demonstrated by President Trump's hiring freeze on biologists for government agencies that largely depend on them, in an attempt to further suppress Science.

2. Of those who try to live intentionally, who make sacrifices great and small, there is understandable outrage and anger, and at times divisive finger-pointing that undercuts efforts to get everyone on the same page. Without doubt this is its own kind of daily trauma that creates a deep wound at the center of authentic efforts and hope.

3. Part of the logic of the Byzantine Christian icon is at once a nod to a past character in Church history, while also being eschatological in that it presents this character loosely, impressionistically, and in projected relation to the resurrection to come.

4. Philip Sherrard, "Human Image, World Image: The Renewal of Sacred Cosmology," in *Towards an Ecology of Transfiguration*, edited by John Chryssavgis and Bruce V. Foltz, (New York: Fordham University Press, 2013), 212.

5. Sherrard, "Human Image, World Image: The Renewal of Sacred Cosmology," 211.

6. Elizabeth Kolbert, in her work, *The Sixth Extinction*, writes that the usual background rate of extinction in ordinary times, according to biologists, is estimated to be .25 percent per million species-years. She writes, "This means that, since there are about fifty-five hundred mammal species wandering around today, at the background extinction rate you'd expect one species to disappear every seven hundred years." Elizabeth Kolbert, *The Sixth Extinction: An Unnatural History* (New York: Henry Holt and Company, 2014), 15.

The story of amphibians, known to be as old as dinosaurs, brings the current rate into startling relief as the group's extinction rate is nearing forty-five thousand times higher than the background rate. As biologist Joseph Mendelson wrote: "I sought a career in herpetology because I enjoy working with animals...I did not anticipate that it would come to resemble paleontology." Ibid., 17.

In his work, *The Open Wound,* Frank Seeburger writes, "The going global of the 'market' economy, brings about the vanishing of the political into the economic. Contrary to the Marxist hope for the withering away of the state, in the global market it is politics that withers away, as the state grows ever more stately." Frank Seeburger, *The Open Wound: Trauma, Identity and Community (*CreateSpace Independent Publishing Platform, 2012*),* 255.

7. Institutional commodification is an enormously complicated question, answered differently in each culture, within the institutions of each culture, and beyond the scope of this paper. But generally, once common land is no longer seen as territory, where forbearers are buried and life entwined, but has instead become property we have entered a de facto negation of aesthetics and opened the path to its destruction. In America, the bottom line is clear: commodification overrules aesthetics.

For instance, the USFS has a mandate to manage the national forests under the principle of "multiple use." But repeatedly, under both legal deposition and policy objectives, they have been unable to quantify the "value" of wilderness solitude, or of seeing a beaver in its environ or a rare mushroom, etc. After all, how do you put a market value on such seeing? Multi-use value, as often happens, is merely the eventual parceling of each thing within a hierarchy of market value. The Forest Service and the BLM never saw a tree they couldn't put a price tag on as a good. In the shifting lexicon of values (natural treasures becoming natural resources, etc.) we lose sight of the goal of the protection of nature, including protecting ourselves as a piece of it. We narrow the notion to an even more anthropocentric notion of the commons, as with human health, and even there the political continues to be usurped by the market. It now seems like a quaint piece of quaint history to look at the EPA'S original Intent of its own goals: Act: SEC. 101 [33 U.S.C. 1251] Declaration of Goals and Policy:

"(a) The objective of this Act is to restore and maintain the chemical, physical, and biological integrity of the Nation's waters. In order to achieve this objective it is hereby declared that, consistent with the provisions of this Act— (1) it is the national goal that *the discharge of pollutants into the navigable waters be eliminated by 1985...*"

8. Bruce V. Foltz, *The Noetics of Nature: Environmental Philosophy and the Holy Beauty of the Visible* (New York: Fordham University Press, 2014), 53.

9. Even the veracity the empirical claim of objectivity is largely debatable, but for our purposes we will go with the dominant narrative. The cliché within so-called "disinterested scientific research" would be important to challenge, however, in a different article. For now, Philip Sherrard's comments taking aim at such a narrative will suffice: "Every thought, every observation, every judgment, every description whether of the modern scientist or of anyone else is soaked in a priori, preconceived, built-in value judgments, assumptions, and dogmas at least as rigid, if not more rigid (because they are so often unconsciously embraced), than those of any explicit religious system." Sherrard, "Human Image, World Image: The Renewal of Sacred Cosmology," 221.

10. This is another reason that the presumption to "valueless" judgment is so naïve. To only grasp one approach to reality assumes that that approach *is* reality. Even if one says that there may be more to it, but we can all agree on this approach, the problem then is the assumption that the "something more" has no impact upon the empirical.

11. Dietrich Bonhoeffer, *The Cost of Discipleship* (New York: Simon & Schuster, 1995), 44-5.

12. Foltz, *Noetics,* 52-53.

13. It is intentional here to use the phrase, "liberation within trauma" instead of "beyond" trauma. To use a religious analogy, Jacob finds liberation after his wrestling match with the angel, but walks away with a limp.

14. We have used Freud's theory of consciousness to mark both a point of origin and likewise to disclaim a universal theory of consciousness formation. The alienation and trauma of that outside world particularly fits with the onslaught of industrialization whereby nature changed in its means of appropriation, such that subject/object differentiation, sadly, became inevitable. Certainly, there are enough accounts from hunter/gather societies that do not suffer such dislocation from nature, nor its resulting issue of belatedness within perception and action like we do, and for whom a much more fluid means of living and dwelling within nature has been documented.

15. Sigmund Freud, *Beyond the Pleasure Principle* (New York: Dover, 2015), 20.

16. Seeburger, *The Open Wound*, 14.

17. Sigmund Freud, *Civilization and its Discontents* (New York: W.W. Norton and Company, 1961), 31. The imagination is discussed as a place of evasion from reality, rather than mediating, which finds its satisfaction in illusions. "The region from which these illusions arise is the life of the imagination; at the time when development of the sense of reality took place, this region was expressly exempted from the demands of reality-testing and was set apart for the purpose of fulfilling wishes where difficult to carry out." As for beauty, he writes, "Psychoanalysis, unfortunately, has scarcely anything to say about beauty either. All that seems certain is its derivation from the field of sexual feeling. The love of beauty seems a perfect example of an impulse inhibited in its aim. Beauty and attraction are originally attributes of the sexual object." Ibid., 34.

18. Frank Seeburger, *The Open Wound,* 14.

19. Freud, *Civilization and Its Discontents*, 25.

20. Ibid., 26

21. This experience, Freud called "traumatic neurosis."

22. Cathy Caruth, *Unclaimed Experience: Trauma, Narrative and History* (Baltimore: Johns Hopkins University Press, 2016), 4. "(T)rauma seems to be more than a pathology, or the simple illness of a wounded psyche: it is always the story of a wound that cries out, that addresses us in the attempt to tell us of a reality or truth that is not otherwise available."

23. Albert Camus, *Carnets: 1935-1942* (New York: Hamish Hamilton Ltd. and Alfred A. Knopf, Inc., 1963), 6.

24. Caruth, *Unclaimed Experience*, 2.

25. Freud, *Civilization and its Discontents*, 33.

26. It is because of such armor that actual repetition, in the Kierkegaardian sense, fails to occur.

27. Robert Duncan, "Often I am Permitted to Return to a Meadow," in *The Opening of the Field* (New York: New Directions, 1960), 7.

28. Jorie Graham, "The Age of Reason," in *Erosion* (New Jersey: Princeton University Press, 1983), 19.

29. Avivah Gottlieb Zornberg, *The Particulars of Rapture: Reflections on Exodos* (New York: Random House, Inc., 2002), 482.

30. Caruth, *Unclaimed Experience*, 7.

31. Ibid.

32. Fyodor Dostoevsky, *The Brothe rs Karam a zov* (New York: Dover, 2005), 291.

33. Zornberg, *The Particulars of Rapture,* 482.

34. Ibid., 423.

35. Seeburger, *The Open Wound.* As Frank Seeburger writes in the first chapter of this present volume, concerning Paul Celan's discussion of the pure poem, "But discourse in the pure poem calls attention to itself for the sake of vanishing altogether before that to which it wants to give voice."

36. Simone Weil, *Gravity and Grace* (New York: Routledge Classics, 2002), 149.

37. Zornberg, *The Particulars of Rapture*, 482.

38. Timothy Morton, *Ecology without Nature: Rethinking Environmental Aesthetics* (Cambridge: Harvard University Press, 2007), 141.

39. Stephen A. Mitchell, *Hope and Dread in Psychoanalysis* (New York: Basic Books, 1993), 102.

40. Morton, *Ecology without Nature,* 141. Morton references Heidegger's discussion in "Building, Dwelling, Thinking," where "thing" comes from the older definition of 'meeting place.'"

41. Zornberg, *The Particulars of Rapture*, 480.

42. Ibid., 481.

43. Ibid., 482.

44. Anne Carson, in *Eros the Bittersweet* explains its etymology: "The Greek word eros denotes 'want,' 'lack,' 'desire for that which is missing.' The lover wants what he does not have. It is impossible for him to have what he wants if, as soon as it is had, he is no longer wanting...desire can only be for what is lacking, not at hand, not present, not in one's possession nor in one's being: eros entails edneia." Anne Carson, *Eros the Bittersweet: An Essay* (New Jersey: Princeton University Press, 1986), 10.

45. Timothy Morton, *Ecology without Nature*, 141.

46. Jeremiah 31:33.

47. Exodus 32:11.

48. Zornberg, *The Particulars of Rapture*, 416-17.

49. Ibid., 498.

50. Simone Weil, *Waiting on God*, (New York: Routledge Revivals, 2009), 61.

51. Carson, *Eros the Bittersweet*, 7-8.

52. These lines come from the movie *Hiroshima mon amour.* They are quoted and discussed by Caruth in *Unclaimed Experience*, 53.

53. Carson, *Eros the Bittersweet*, 11.

54. This is demonstrated by the sheer surprise of Moses as he comes down to see the people have made, in his absence, a golden calf.

55. Zornberg, *The Particulars of Rapture*, 421.

56. Zornberg, *The Particulars of Rapture*, 498.

57. St. Vladimir's Theological Quarterly, Journal for the Study of Religion, Nature and Culture, Journal of the American Academy of Religion. https://www.pdcnet.org/enviroethics, https://www.pdcnet.org/envirophil/Environmental-Philosophy,

58. F. Scott Fitzgerald, *The Great Gatsby* (New York: Simon and Schuster, 2003), 190. Also see Megan McGilchrist, *The Western Landscape in Cormac McCarthy and Wallace Stegner; Myths of the Frontier (New York: Routledge, 2010),* 163.

59. Kolbert, *The Sixth Extinction,* 27. Also see *Old World, New World: America and Europe in the Age of Jefferson*, edited by Leonard Sadosky, (University of Virginia Press, 2010).

60. Kolbert, *The Sixth Extinction,* 27-37. Also see Thomas Jefferson, *Notes on the State of Virginia* (Boston: Lilly and Wait, 1832), 52.

61. John James Audubon, *The Missouri River Journals of John James Audubon* (Lincoln: University of Nebraska Press, 2016), 61.

62. Anne E. Maczulak, *Conservation: Protecting Our Plant Resources* (New York: Facts on File, 2010), xiv.

63. Lee A. Vedder, *John James Audubon and the Birds of America: A Visionary Achievement in Ornithological Illustration* (San Marino: Henry E. Huntington Library and Art Gallery, 2006), 6.

64. Nancy Plain. *This Strange Wilderness: The Life and Art of John James Audubon* (Lincoln: University of Nebraska Press, 2015).

65. Rev. D. Mead, ed. "John James Audubon" in *The American Literary Emporium Or Friendship's Gift* (New York: C.H. Camp, 1848), 209.

66. Zornberg, *The Particulars of Rapture*, 482.

67. Ibid.

68. Zornberg, *Particulars of Rapture*, 480.

69. Ibid.

70. There are other methods of showing animation and scale. In China, for instance, and especially before the Ming Dynasty, there has been a long relationship, dating from the third and fourth century of poets and painters working together. The copying of the living bird into the art was preferred to killing the bird to get its exact representation. See David Hinton, *Existence: A Story* (Boulder: Shambhala, 2016).

71. Of the men he travelled with, Audubon confessed, "The Fur company may be called the exterminating medium of these wild and almost uninhabitable regions, which cupidity or the love of money alone would induce man to venture. Where can I go now and find nature undisturbed?" Maria R. Audubon, with Zoological and Other Notes by Elliott Coues, "The Labrador Journal" in *Audubon and His Journals Volume One* (New York: Charles Scribner's Sons, 1897), 379.

72. We are not trying to say that aesthetics of violence does not have its place, as it most certainly does, and some of the most powerful art is drawn from this, but rather here we are speaking to its ethics as being inseparable from intent of production.

73. Cathy Caruth, *Unclaimed Experience*, 107.

74. Ibid., 107.

75. Shuli Barzilai, *Lacan and the Matters of Origin,* (Stanford University Press, 1999), 39-40. Barzilai argues for the early, nearly impossible tension, between love and a notion of a complete self. For a discussion that is slightly more optimistic on the capacity of love within the Mirror Stage, see Zornberg, *The Particulars of Rapture,* 61-72.

76. For a discussion on Freud's idea and speculation that consciousness is founded on an awakening from death, see Cathy Caruth, *Unclaimed Experience,* 106-9.

77. Zornberg, *The Particulars of Rapture,* 65.

78. Caruth, Unclaimed Experience, 99-100. Also see Seeburger, *Open Wound*, 14-17.

79. Caruth, *Unclaimed Experience,* 100.

80. Jorie Graham, *Place* (New York: HarperCollins, 2012), 5.

81. Foltz in *The Noetics of Nature* notes that Heidegger draws this out even farther, suggesting that "Being itself somehow presents us with a face, and ultimately refers to us not to an "itself" at all but to a "Thou" who demands that we pass from ontology and aesthetics to theology." Foltz, *The Noetics of Nature*, 141.

82. There is a long history of the interpretation of the dream of a burning child. For Freud, the dream tempts consciousness to go on sleeping. For Lacan, dream will remain the site of ethical training wherein the fire of loss is felt inside and outside. For Zizek, this dream is not the site of fiction, but where consciousness breaks out of its ideology, into the real. For our purposes, it will be enough to show that the dream is ethical and edifying in its orientation.

83. Sigmund Freud, *The Interpretation of Dreams*, (New York: Macmillan, 1913), 403.

84. Ibid., 404.

85. Caruth, *Unclaimed Experience*, 103.

86. As with Moses, it is impossible to say where the reality of the child ends or begins with the father, whether in the dream or awake. It shows a communion that is more intimate than consciously experienced.

87. Seeburger, *Open Wound*, 17.

88. Fydor Dostoevsky, *Demons*, translation by Richard Pevear and Larissa Volokhonsky (New York: Vintage Classics, 1994), 255.

89. Seeburger, *The Open Wound,* 17. The call to stay awake litter the Gospels, from the maiden women and their lamps, to the Apostles themselves who fall asleep on their watch.

90. Ibid.

91. Ibid.

92. Ibid.

93. For elaboration, see Slavoj Zizek, *The Sublime of Ideology* (New York: Verso, 1989).

94. Caruth, *Unclaimed Experience*, 103.

95. Foltz, *The Noetics of Nature*, 157.

96. Zizek's position of the gap fits into this idea of epiphanic gaps. As Caruth writes, pg. 170, of Unclaimed Experience: "Slavoj Zizek suggests that the awakening in Lacan's reading of the dream is a precise reversal of the usual understanding of dream as fiction and of awakening as reality: he argues that the awakening of the father in Lacan's reading is an "escape" from the real into ideology. " She goes onto sum, "the encounter with the real cannot simply be located either inside or outside the dream, but has to be located in the moment of the transition between the two, in the movement from one to the other. This is what Lacan precisely calls, "the gap that constitutes the awakening." Caruth, *Unclaimed Experience*, 170.

About the Authors

Joshua Coleman, PhD, teaches philosophy and world religions to seniors at St. Michael's Catholic Academy in Austin, TX. He has also taught social justice at the same school. Dr. Coleman is currently publishing an article on philosophy and the music group The Band, as well as writing a book on the relationship between religion and college football in the Deep South.

Sharon Mar Adams graduated from the University of Denver and Iliff School of Theology Joint PhD program with a concentration in theology, philosophy and cultural theory. She has taught courses in religion, women and gender studies, philosophy and the humanities at the University of Colorado at Boulder. She received the Vinnik Fellowship in order to study religious art and interfaith dialogue in Israel.

Mark Flory is an instructor at Metropolitan State University of Denver and the Community College of Denver. He teaches philosophy, religion, and mythology. Mark's research focuses primarily on comparative spiritual practices, the theory of practice, the phenomenology of spiritual progress, and related matters.

J. Porter holds an MA in philosophy from the University of Denver and an MFA from the Iowa Writer's Workshop. She is currently working on two novels and two screenplays.

Frank Seeburger, PhD, is Emeritus Professor of Philosophy at the University of Denver. He is the author of *Addiction and Responsibility: An Inquiry into the Addictive Mind* and *The Open Wound: Trauma and Community,* as well as various other books and articles. For a number of years, his work has focused on issues concerning the intersection of trauma and philosophy, on which topic he has a blog available at traumaandphilosophy.com.

Donald L. Turner teaches philosophy and religion courses at Nashville State Community College. His work appears in journals such as *Philosophia* and *Disclosure* and in collections such as *Heidegger and the Earth: Issues in Environmental Philosophy* (Toronto University Press, 2009), *Beyond Human: From Animality to Transhumanism* (Continuum, 2012), and *Models of God and Other Ultimate Realities* (Springer, 2013). His research and writing focus on twentieth-century Continental philosophy and questions about ethics, religion, and animality.